Setting Priorities for Drinking Water Contaminants

Committee on Drinking Water Contaminants

Water Science and Technology Board

Board on Environmental Studies and Toxicology

Commission on Geosciences, Environment, and Resources

NATIONAL ACADEMY PRESS
Washington, D.C. 1999

Support for this project was provided by the Environmental Protection Agency under contract no. X-826345-01-0.

Library of Congress Catalog Card Number 98-88191

International Standard Book Number 0-309-06293-4

Additional copies of this report are available from:

National Academy Press
2101 Constitution Avenue, NW
Box 285
Washington, D.C. 20055
800-624-6242
202-334-3313 (in the Washington metropolitan area)
http://www.nap.edu

Preface

Americans expect to be able to drink the water that comes from their taps without fear for their safety. While this expectation has been largely fulfilled in this century, on occasions and in certain systems, chemicals and microbes still contaminate drinking water supplies in the United States. Thus, continuing vigilance is necessary to assure that important drinking water contaminants are identified and appropriately addressed.

In 1996, Congress enacted the Safe Drinking Water Act Amendments of 1996, substantially revising the way in which public water supplies are regulated. Among the amendments is a requirement that the Environmental Protection Agency (EPA) develop a list of currently unregulated contaminants that may pose risks in drinking water. Every five years, EPA must decide whether to regulate at least five of those contaminants and whether to conduct additional monitoring and research for the others. The first of these lists, known as the Drinking Water Contaminant Candidate List (CCL), was published by EPA in March of 1998.

EPA has asked the National Research Council (NRC) for assistance in addressing three aspects of this effort:

1. developing a scientifically sound approach for deciding whether or not to regulate contaminants on the current and future CCLs,
2. convening a workshop that will focus on emerging drinking water contaminants and the database that should be created to support future decision making on such contaminants, and
3. creating a scientifically sound approach for developing future CCLs.

This report, written by the NRC's Committee on Drinking Water Contaminants, addresses the first of these topics. It includes a review of ten different governmental and private schemes used to rank environmental contaminants in other contexts to identify whether, and in what manner, such schemes might apply to the three tasks being addressed by the committee. The committee consists of 14 volunteer experts in water treatment engineering, water chemistry, analytical chemistry, microbiology, toxicology, public health, epidemiology, risk assessment, and risk communication. This report's findings are based on a review of relevant technical literature, information gathered at two committee meetings, and the expertise of committee members. The committee will address the remaining two topics in subsequent reports.

On behalf of the committee, I wish to thank Jim Taft, Evelyn Washington, and Chuck Job of EPA for supporting this important three-phased review. In addition, thanks go to William Carpenter, attorney and member of the National Drinking Water Advisory Council's Working Group on Occurrence and Contaminant Selection; Stephen Clark, EPA Office of Ground Water and Drinking Water; and Thomas Yohe, Philadelphia Suburban Water Company. All of these individuals made presentations and supplied valuable background information at the first committee meeting.

This report has been reviewed, in accordance with NRC procedures, by individuals chosen for their expertise and broad perspectives on the issues addressed herein. The purpose of this review is to provide independent, candid, and critical comments that will help the NRC to be assured that the report is sound and meets the NRC's standards for objectivity, evidence, and responsiveness to the study charge. The review comments and draft report remain confidential to protect the deliberative process by which the report was developed. The committee wishes to thank the following people for their participation in this review and for their many constructive comments: Richard Bull, Washington State University and Battelle Pacific Northwest Laboratory; Gunther Craun, Gunther F. Craun & Associates, Staunton, Virginia; Joseph Delfino, University of Florida; George Hornberger, University of Virginia; Eric Olson, National Resources Defense Council; David Reckhow, University of Massachusetts, Amherst; Robert Spear, University of California, Berkeley; and Thomas Yohe, Philadelphia Suburban Water Company. Notwithstanding this review, however, the final content of this report is the responsibility of the Committee on Drinking Water Contaminants.

I speak for the whole committee in thanking the very capable and professional NRC staff for the assistance we have received throughout our deliberations and during report preparation. In particular, I want to acknowledge the outstanding efforts we have received from

Jacqueline MacDonald, study director and associate director, Water Science and Technology Board (WSTB),

Carol Maczka, director of toxicology and risk assessment programs, Board on Environmental Studies and Toxicology,

Mark Gibson, research associate, WSTB, and

Kimberly Swartz, project assistant, WSTB.

These staff members worked extraordinarily hard and effectively to help us produce this report in a very short time, in order to be of maximum utility to EPA as it moves forward to implement the recently enacted drinking water protection mandates.

Finally, I want to thank the diverse and talented members of the committee, who were able to bring together their diverse perspectives and broad expertise to produce this report. I look forward to continuing to work with this wonderful group in making a contribution to addressing the second and third phases of our effort.

Warren R. Muir, Ph.D.
Chair, Committee on Drinking Water Contaminants

The National Academy of Sciences is a private, nonprofit, self-perpetuating society of distinguished scholars engaged in scientific and engineering research, dedicated to the furtherance of science and technology and to their use for the general welfare. Upon the authority of the charter granted to it by the Congress in 1863, the Academy has a mandate that requires it to advise the federal government on scientific and technical matters. Dr. Bruce M. Alberts is president of the National Academy of Sciences.

The National Academy of Engineering was established in 1964, under the charter of the National Academy of Sciences, as a parallel organization of outstanding engineers. It is autonomous in its administration and in the selection of its members, sharing with the National Academy of Sciences the responsibility for advising the federal government. The National Academy of Engineering also sponsors engineering programs aimed at meeting national needs, encourages education and research, and recognizes the superior achievements of engineers. Dr. William A. Wulf is president of the National Academy of Engineering.

The Institute of Medicine was established in 1970 by the National Academy of Sciences to secure the services of eminent members of appropriate professions in the examination of policy matters pertaining to the health of the public. The Institute acts under the responsibility given to the National Academy of Sciences by its congressional charter to be an adviser to the federal government and, upon its own initiative, to identify issues of medical care, research, and education. Dr. Kenneth I. Shine is president of the Institute of Medicine.

The National Research Council was organized by the National Academy of Sciences in 1916 to associate the broad community of science and technology with the Academy's purposes of furthering knowledge and advising the federal government. Functioning in accordance with general policies determined by the Academy, the Council has become the principal operating agency of both the National Academy of Sciences and the National Academy of Engineering in providing services to the government, the public, and the scientific and engineering communities. The Council is administered jointly by both Academies and the Institute of Medicine. Dr. Bruce M. Alberts and Dr. William A. Wulf are chairman and vice-chairman, respectively, of the National Research Council.

Contents

Setting Priorities for Drinking Water Contaminants

Executive Summary

The provision of safe drinking water has been a major triumph of twentieth-century U.S. public health practice and has been an important factor in the improvement of the health status of U.S. communities since the turn of the last century. Nevertheless, chemical and microbiological contaminants still occur in drinking water supplies. Further, waterborne disease has not been entirely eliminated in the United States, as was evident in the major cryptosporidiosis outbreak that affected some 400,000 Milwaukee residents in 1993. The continuing presence of contaminants in water supplies and occurrences of waterborne disease serve as reminders that the system for regulating drinking water in the United States needs to be reassessed periodically.

The most recent update of U.S. policies for ensuring the safety of public water supplies occurred in the summer of 1996, when Congress enacted the Safe Drinking Water Act (SDWA) Amendments of 1996. These amendments substantially revised the process for regulating public water supplies. One of the major changes was the creation of a new policy for establishing standards for contaminants that currently are unregulated. The amendments require that every five years, the Environmental Protection Agency (EPA) develop a list of currently unregulated contaminants that may pose risks in drinking water and decide whether to regulate at least five of those contaminants. In March 1998, EPA published the first of these lists, known as the Drinking Water Contaminant Candidate List (CCL), of priority unregulated contaminants. Subsequently, EPA asked the National Research Council (NRC) for assistance in developing a process for setting priorities among the listed contaminants. This report responds to that request. Specifically, the report evaluates various existing schemes for set-

1

ting priorities among environmental contaminants and recommends a framework to guide EPA in deciding which contaminants on the CCL to regulate, which to monitor, and which to study further.

This report was written by the NRC's Committee on Drinking Water Contaminants. The committee was appointed in 1998 in response to EPA's request for assistance in establishing a priority-setting process for drinking water contaminants. EPA also requested help in establishing processes for identifying emerging drinking water contaminants and developing future CCLs. The committee consists of 14 volunteer experts in water treatment engineering, water chemistry, analytical chemistry, microbiology, toxicology, public health, epidemiology, risk assessment, and risk communication. This report's findings are based on a review of relevant technical literature, information gathered at two committee meetings, and the expertise of committee members. Future committee reports will advise EPA on emerging drinking water contaminants and establishment of future CCLs.

In reviewing this report, the reader should keep in mind that the committee was guided first and foremost by concerns about public health. The committee chose this perspective because public health is the basis for the SDWA; the SDWA directs the EPA administrator to consider "contaminants for listing that, first, may have an adverse effect on the health of persons." Further, the report takes the position that scientific disagreements about the public health effects of contaminants and their relative severity are the norm and do not signal a deviation from sound science. Paradoxically, when data are sparse they often appear consistent and coherent (for example, when produced by one or a few laboratories), but data gaps become evident as the problem is examined more fully by different methods and from different perspectives. The EPA faces a challenging task in assessing the available scientific information about contaminant risks and, based on that assessment, making a risk management decision about which contaminants should be regulated. In this process, there is no replacement for policy judgments by EPA. The committee purposely declined to define what constitutes "sufficient" data for making decisions related to drinking water contaminants, because this is a matter of judgment that will vary with context.

EXISTING PRIORITIZATION SCHEMES

Government agencies and private industries have developed a number of schemes that rank chemicals according to their importance as environmental contaminants. No equivalent schemes exist for microbial contaminants. The committee reviewed ten chemical prioritization schemes (see Table ES-1) to examine what methodological elements and data considerations in these schemes may be useful for prioritizing drinking water contaminants. All the schemes prioritize chemicals on the basis of risk to human health and/or the environment

TABLE ES-1   Chemical Prioritization Schemes Reviewed in This Study

Contaminant Prioritization Schemes Reviewed	Source[a]	Contaminant Prioritization Function
Cadmus Risk Index Approach	Cadmus Group	Drinking water contaminants
American Water Works Association (AWWA) Screening Process	AWWA	Drinking water contaminants
Proposed Regulation Development Process	AWWA, National Association of Water Companies (NAWC), Association of Metropolitan Water Agencies (AMWC), and Association of State Drinking Water Administrators (ASDWA)	Drinking water contaminants
Waste Minimization Prioritization Tool	EPA Office of Solid Waste and Emergency Response and Office of Pollution Prevention and Toxics	All potential environmental contaminants
Section 4(e) of Toxics Substances Control Act	Interagency Testing Committee	All potential environmental contaminants
State of California Safe Drinking Water and Toxic Enforcement Act of 1986 (Proposition 65)	California Environmental Protection Agency	All potential environmental contaminants
Hazard Ranking System (HRS)	EPA	Hazardous waste sites
Comprehensive Environmental Response, Compensation, and Liability Act Priority List of Hazardous Substances	Agency for Toxic Substances and Disease Registry and EPA	Hazardous materials
Sediment Contaminant Inventory Hazard Analysis of Releases Inventory	EPA Office of Science and Technology	Sediment contaminants
Pesticide Leaching Potential (PLP)	EPA Office of Pesticide Programs	Pesticides

[a]Agency, industry, or act responsible for the development of the ranking scheme.

by considering exposure and toxicity. Each scheme is unique, however, in its use of data and ranking criteria.

The committee concluded that a ranking process that attempts to sort contaminants in a specific order is not appropriate for the selection of drinking water contaminants from the CCL for regulation. In the absence of complete information, the output of prioritization schemes is so uncertain (though this uncertainty is generally not stated) that they are of limited use in making more than preliminary risk management decisions about drinking water contaminants. While the ranking schemes the committee evaluated presumably are useful for their intended purposes and may provide a quantitative means for screening and sorting large numbers of contaminants, their accuracy is not sufficient for prioritizing a relatively small number of contaminants, many of which may pose similar degrees of risk for drinking water. Simple quantitative ranking processes, such as those the committee examined, cannot substitute for policy judgments by EPA when moving toward final regulatory decisions. Furthermore, if enough information is available to determine that a contaminant occurs in drinking water at levels and frequencies that may pose a health risk, then the contaminant should be considered for regulation, without attempting to assign a priority to it. That said, however, the existing methods for ranking environmental contaminants may be useful for sorting large numbers of potential contaminants not yet on the CCL to determine which ones should be included on future CCLs. The committee will use its analysis of these schemes in providing guidance on CCL development in a future report.

THE 1998 CCL

The existing CCL is essentially an unprioritized list of research needs for drinking water contaminants. Additional research will need to be conducted for many, if not most, of the contaminants on the current CCL. At the same time, however, to be included on the list contaminants had to pass a relatively rigorous screening process involving assessment of existing contaminant occurrence and heath effects data by a group of experts (the National Drinking Water Advisory Council Working Group on Occurrence and Contaminant Selection). A variety of stakeholders, including representatives of the water utility industry and public interest groups, commented on the list, and the list was revised accordingly. Thus, the contaminants on the current CCL may have a much higher likelihood of posing risks in drinking water than would a randomly assembled list of unregulated chemicals and microorganisms. A key question for EPA is how to determine which contaminants can be moved off this research list and into the regulatory arena. Ideally, EPA will develop a process for making these determinations that will apply not only to the current CCL but also to future CCLs.

SELECTING CONTAMINANTS ON THE CCL FOR FUTURE ACTION

The committee recommends that EPA use a phased process, shown in simplified outline in Figure ES-1, for determining which contaminants on the CCL are appropriate candidates for regulatory action and which will require research. Figure ES-1 includes a recommended time line for completing each phase of the process. The time line is provided to help EPA allocate time and resources in order to meet the 1996 SDWA Amendments' requirement to publish regulatory determinations (i.e., regulate or not regulate) for at least five contaminants on the CCL by August 2001. However, the committee recognizes that almost one year of the originally allotted time (i.e., three and one-half years following publication of the first CCL) have already passed. Thus, the time line should not be strictly interpreted, but should serve as a guide for the relative amount of time to allocate to each step in the process. The suggested time line should be of more direct use following the publication of future CCLs.

Chapter 5 explains the details of the decision process shown in Figure ES-1. In brief outline, the process would proceed as follows:

• Within approximately one year of completion of the CCL, EPA should conduct a three-part assessment of each contaminant on the CCL. For each contaminant, the three parts consist of (1) a review of existing health effects data, (2) a review of existing exposure data, and (3) a review of existing data on treatment options and analytical methods. The first part of the assessment should consider data on the contaminant's effects on vulnerable subpopulations such as pregnant women, infants, the elderly, and those with compromised immune systems. While general guidelines for reviewing existing data are possible and are presented in Chapter 5, an important component of the reviews will be policy judgments by EPA about the significance of the data.

• After completion of the three-part assessment, EPA should conduct a preliminary risk assessment based on available data identified in the three-part assessment. The risk assessment, which integrates hazard and exposure analyses to estimate the public health implications of the contaminant, should be carried out even if there are data gaps; this will provide a basis for an initial decision about the disposition of the contaminant and guide research efforts, where needed. The preliminary risk assessment should not be overly detailed or resource intensive. EPA's usual approach to risk assessment is appropriate, and the committee does not see a need to create new procedures for this step.

• After completing the preliminary risk assessment, EPA should prepare a separate decision document for each contaminant that indicates whether the contaminant will be dropped from the CCL because it does not pose a risk, will be slated for additional research (e.g., on health effects, exposure, or risk reduction), or will be considered for regulation. The decision document should explain the reasoning for EPA's determination and should be publicly disseminated for com-

6

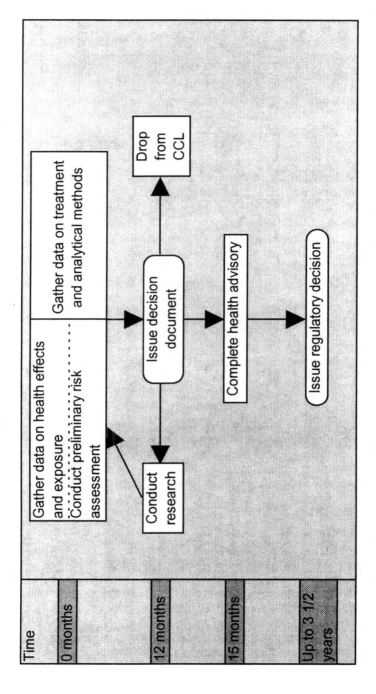

FIGURE ES-1   Phased process for setting priorities among contaminants on the CCL.

ment. Decision documents for contaminants dropped from the CCL should specify the health and exposure data that EPA used to conclude that the contaminant poses little or no risk.

• When the three-part assessment or preliminary risk assessment identifies important information gaps, EPA should develop a research and monitoring plan to fill such gaps in time to serve as the basis for a revised assessment and decision document before the end of the three-and-a-half-year cycle required by Congress for evaluating contaminants on the CCL. In filling information gaps, EPA should solicit the voluntary participation of industry and others and should use its other authorities (such as those under the Toxic Substances Control Act and the Federal Insecticide, Fungicide, and Rodenticide Act) to help fill data gaps.

• Health advisories should be issued for all contaminants remaining on the CCL after completion of an initial set of decision documents. A health advisory is an informal technical guidance document that defines a nonregulatory (i.e., nonenforceable) concentration of a drinking water contaminant for which no adverse health effects would be anticipated over specific exposure durations. To provide the public with the best available information about the contaminant, EPA should develop a health advisory for any contaminant for which credible evidence of a risk in drinking water exists, even if existing data are insufficient to develop a full regulation. Contaminants subject to a health advisory may need additional research and monitoring even after completion of a revised assessment and decision document.

As indicated in Figure ES-1, decisions to drop a contaminant from the CCL, to issue a health advisory, or to proceed toward regulation should be based on health risk considerations only. However, EPA should fill data gaps in treatment technologies and analytical methods to avoid delaying regulatory action for contaminants for which current information on treatment and detection is inadequate.

In implementing this phased process, EPA should keep in mind that it should act immediately on any contaminant that meets the statutory tests of (1) a determination that the contaminant "may" adversely affect public health, (2) evidence that the contaminant is known or substantially likely to occur in public water systems with a frequency and at levels that pose a threat to public health, and (3) an indication that controlling the contaminant presents a meaningful opportunity for health risk reduction. Development of regulations for contaminants that meet these three requirements (which are specified in the SDWA amendments) should not be delayed by implementation of the phased approach. The ability to act quickly and short circuit the phased evaluation process is especially critical for newly discovered high-risk contaminants.

EPA should also keep in mind that a mechanistic, quantitative ranking approach—a system that would mechanically process all information about contaminants and produce a regulatory determination — is not appropriate for prioritizing contaminants on the CCL. While Chapter 5 of this report provides details about data and criteria that EPA should consider in its three-part reviews of contaminants, it does not specify a mechanistic tool (one free of policy judgments) for assessing contaminants. The need for policy judgments by EPA cannot and should not be removed from this process. Ultimately, EPA is accountable to the public for the decisions it makes about regulating drinking water contaminants. In making these decisions, EPA should use common sense as a guide and should err on the side of public health protection.

1

Introduction

The establishment of laws, advisories, and standards to protect public drinking water supplies has been a major endeavor at the federal, state, and local government levels for decades. The provision of safe drinking water has been one of the major triumphs of twentieth century U.S. public health practice and has been a major factor in the improvement of the health status of U.S. communities.

For most Americans, in this century turning the tap has been an act of faith in the generally safe character of our public water supplies. The increasing consumption of bottled waters, whose cost is hundreds of times the cost of tap water, however, suggests that the public has begun to question that faith. At the same time, despite the presence of several layers of regulatory protection, many sources of raw and finished public drinking water in the United States contain chemical, microbiological, and even radiological contaminants at detectable levels (Neal, 1985). Some of these contaminants pose risks not only via ingestion of the contaminated water (which can be avoided by drinking bottled water) but also via dermal contact with the water or inhalation of vapors while showering. A recent National Research Council (NRC) report, *Safe Water From Every Tap* (NRC, 1997), noted that 23.5 percent of all U.S. community public water systems violated microbiological standards under the Safe Drinking Water Act (SDWA) at least once between October 1992 and January 1995 and that 1.3 percent violated chemical standards. The frequent presence of such contaminants, as well as documented outbreaks of waterborne disease and the many other outbreaks thought to go undetected, are a clear reminder that unprotected and contaminated drinking water can still pose health risks to the population.

When Congress amended the SDWA in 1996, it required the U.S. Environmental Protection Agency (EPA) to prepare periodically a list of unregulated contaminants[1] to assist in priority-setting efforts for the agency's drinking water program. Subsequently, the first Drinking Water Contaminant Candidate List (CCL) was published in final form in the *Federal Register* (63 FR 10274) on March 2, 1998 (EPA, 1998a). Under the authority of the SDWA Amendments of 1996, and at the direction of Congress, EPA asked the NRC to recommend criteria for prioritizing CCL contaminants into categories for future regulatory action. EPA will need to classify contaminants into those ready for a rule-making decision; those ready for guidance development, including health advisories;[2] those needing additional occurrence[3] data; and those that are priorities for additional health effects or other research.

Under the SDWA amendments, EPA must determine whether or not to regulate at least five contaminants on the CCL by August 2001 (three and one-half years following publication of the first CCL). To support these decisions and meet the statutory language of the SDWA amendments, EPA must determine whether the regulation of each contaminant on the CCL would provide a "meaningful opportunity" to reduce health risk for persons served by public water systems (EPA, 1998b). EPA must also consider risk to sensitive subpopulations such as infants and the elderly (EPA, 1997a). To help meet these requirements, EPA requested recommendations from the NRC for a process or criteria to help select contaminants from the CCL for regulatory determinations.

This report addresses EPA's request. It was prepared by the NRC's Committee on Drinking Water Contaminants, appointed in 1998 in response to EPA's solicitation. The committee consists of 14 volunteer experts in water treatment engineering, toxicology, public health, epidemiology, water and analytical chemistry, risk assessment, risk communication, public water system operations, and microbiology. Members convened twice over a six-month period to develop this report. The group incorporated input from a wide range of stakeholders and EPA personnel concerned about the health, economic, and technological implications of contaminants included on the CCL. The committee also sought the input of representatives of the National Drinking Water Advisory Council's (NDWAC's)

---

[1] According to SDWA Section 1401(6), "The term 'contaminant' refers to any physical, chemical, biological, or radiological substance or matter in water." This definition has not been revised since the inception of the SDWA in 1974 and includes nontoxic and potentially beneficial "contaminants."

[2] A health advisory is an informal technical guidance document that defines a nonregulatory (i.e., nonenforceable) concentration of a drinking water contaminant at which no adverse health effects would be anticipated to occur over specific exposure durations, including a margin of safety (EPA, 1996).

[3] For the purposes of this report, the committee broadly defines the term "occurrence" of a contaminant as the presence of a measurable amount of the contaminant in a water supply. In contrast, contaminant "exposure" is defined as human contact with that substance (or its component byproducts) via the potable water supply.

Working Group on Occurrence and Contaminant Selection. This group was integral to the identification and selection of potential drinking water contaminants for inclusion on the 1998 CCL (EPA, 1997a).

This chapter provides a brief overview of the historical development of drinking water regulations, especially the SDWA and its 1996 amendments. It also summarizes the development, purpose, and requirements of the CCL and other related SDWA provisions. Chapter 2 responds to EPA's request for the committee to review and summarize several schemes for prioritizing chemical contaminants according to risk. Chapter 3 briefly summarizes methods used to evaluate risks posed by microbiological contaminants in drinking water. Chapter 4 describes how the 1998 CCL was developed. Lastly, Chapter 5 recommends a decision framework and provides general guidelines for evaluating contaminant-related data for the selection of CCL contaminants for future regulatory action.

HISTORICAL DEVELOPMENT OF WATER SUPPLY REGULATIONS

The essential benefits of filtration and chlorination of potable water supplies were established in the United States by the beginning of World War I (NRC, 1977). The primary reason for water purification was to protect public health from typhoid fever and other waterborne diseases.

Since 1920, waterborne disease outbreaks have been investigated and reported in a national database, which is operated jointly by EPA and the Centers for Disease Control and Prevention (CDC). According to this database, from 1920 to 1970 there were 1,085 waterborne disease outbreaks in the United States (see Figure 1-1). From 1971 to 1992, more than 164,000 individuals were reported ill during 684 documented waterborne outbreaks (Craun, 1991; Herwaldt et al., 1992; Moore et al., 1993). Interestingly, the average number of outbreaks for the first 50 years was 21 per year, and for the last 22 years the average has been 31 per year. Thus, despite improvements in water treatment, documented outbreaks continue to occur. A portion of this reported increase may be because of improved monitoring, detection, and reporting methods. Nevertheless, epidemiologists generally agree that these reported occurrences represent only a fraction of total waterborne disease outbreaks. The national waterborne disease database depends on detection, investigation, and complete reporting by individual states. However, state data are known to be uneven in quality and do not provide information about undetected or undocumented outbreaks. Furthermore, reporting is voluntary. For example, some states support active surveillance and investigation of waterborne disease outbreaks, while others rely on case reports provided by local physicians and public health officers regarding clusters of illness.

Until passage of the original SDWA in 1974, there was no enforceable provision in federal law to protect the public from hazardous chemical substances in drinking water. Prior to the SDWA, the only enforceable federal drinking

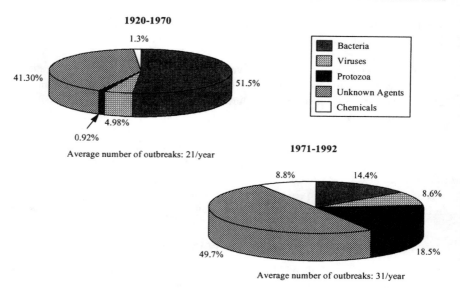

FIGURE 1-1   Causes of waterborne disease outbreaks reported to the CDC.
SOURCES: Adapted from Craun, 1991; Herwaldt et al., 1992; Moore et al., 1993.

water standards were directed at waterborne pathogens in water supplies utilized by interstate carriers such as buses, trains, airplanes, and ships (Viessman and Hammer, 1985; Dzurik, 1990; NRC, 1997). These standards were originally promulgated under the Public Health Service Act of 1912 (PHSA). While the PHSA did include recommended guidelines for drinking water contaminants unrelated to communicable diseases, they were not enforceable (Dzurik, 1990; NRC, 1997).

The purpose of the original SDWA was to ensure that public water supply systems[4] meet national primary drinking water regulations for contaminants to protect public health. The SDWA also established a joint federal-state system for ensuring compliance with federal standards. Since its enactment, the SDWA has been significantly amended twice: first in 1986 and, most recently, in 1996. This report is principally concerned with requirements newly established in the SDWA Amendments of 1996.

---

[4]Under the SDWA Amendments of 1996, distribution systems providing water for human consumption through "constructed conveyances" (e.g., pipe networks, irrigation ditches) to at least 15 service connections or an average of 25 individuals daily at least 60 days per year are defined as public water systems subject to SDWA regulation (EPA, 1998c). According to EPA (1994), there are more than 190,000 public water systems in the United States (including those in U.S. territories and Native American lands).

Under the SDWA, setting a national primary drinking water regulation for a chemical contaminant is a two-step process (Gibson et al., 1997). First, a nonenforceable maximum contaminant level goal (MCLG) is determined. This criterion represents the level in drinking water that would result in "no known or anticipated adverse effect on health" with a margin of safety. Second, MCLGs serve as the target for setting either enforceable national primary drinking water standards, known as maximum contaminant levels (MCLs), or treatment techniques, if contaminant monitoring is not feasible. In general, MCLs are set as close to the MCLG as feasible, depending on risk management considerations (e.g., EPA determines that the cost of a standard at the MCLG is not justified by the benefits) (EPA, 1996a).

For microbiological contaminants, philosophically the original SDWA established a zero tolerance for disease-causing organisms as the health goal (i.e., the MCLG is set at zero). However, treatment performance techniques, rather than specific allowable concentration of pathogens, historically have served as the basis for regulating microbial contaminants. Historically, water supply regulators assumed that all waters carried some level of harmful organisms that could be treated generically, with levels of fecal coliform bacteria (which are generally harmless) serving as an overall measure of the performance of the treatment system. The SDWA Amendments of 1986 and 1996 both required some modifications to this historical approach. The 1986 amendments required development of the Surface Water Treatment Rule to optimize filtration and disinfection of surface waters to protect against microorganisms, such as *Giardia*, that resist treatment and those, such as *Legionella* (cause of Legionnaires' disease) that can grow in the water distribution system. Strong source water protection programs in some cases can supplement or supplant filtration requirements of this rule. The 1996 amendments require that methods be considered to protect the population from exposure to recently recognized waterborne pathogens, such as *Cryptosporidium*.

The NRC helped EPA establish the first set of national primary drinking water regulations for individual contaminants and contaminant classes under the original SDWA (NRC, 1977). The resulting NRC report *Drinking Water and Health* dealt with standards for chemical, microbiological, particulate, and radionuclide drinking water contaminants. NRC is now being asked to assist EPA in its task of identifying and selecting contaminants for future regulatory attention through the mechanism of the 1998 CCL. The EPA originally requested that this study focus primarily on chemical contaminants. After preliminary deliberations, however, the committee decided to pay equal attention to both chemical and microbiological contaminants.

DEVELOPMENT OF THE CCL

The first CCL identification and selection process used separate approaches for microbiological and chemical contaminants. The CCL comprises 60 contaminants and contaminant classes, including 50 chemicals and chemical groups and 10 microbiological contaminants and groups of microbes, as listed in Table 1-1 (EPA, 1997a).

With the exception of sulfate (included as a special case), the CCL includes contaminants that are not currently subject to any proposed or promulgated primary drinking water regulation, but are known or anticipated to occur in public water systems and may require regulation under the SDWA. Thus, the 1998 CCL is intended to be the primary source of priority contaminants for future regulatory actions by EPA's drinking water program until the next CCL is published in 2003 (see Figure 1-2).

RELATED SDWA PROGRAMS

As indicated in Figure 1-2, future CCL development will be closely interrelated with two other drinking water programs established by the SDWA Amendments of 1996: the National Drinking Water Contaminant Occurrence Database (NCOD) and the Unregulated Contaminant Monitoring Regulation (UCMR) (EPA, 1998b). Both of these programs, as well as the CCL, are currently the responsibility of EPA's Office of Ground Water and Drinking Water.

The purpose of the NCOD is to store quality-controlled data on the occurrence of regulated and unregulated drinking water contaminants (EPA, 1998b). When operational, the NCOD is expected to provide the basis for identifying drinking water contaminants that may be included on future CCLs and to support EPA's decisions about whether to regulate contaminants in the future. It is also expected to be used in the periodic review of existing contaminant regulations and monitoring requirements. The NCOD is currently under development and, by law, will need to be completed by August 1999. EPA has requested input from the public, states, and the scientific community regarding the NCOD's design, structure, and use (AWWA, 1997; EPA, 1997b).

The 1996 SDWA Amendments also direct EPA to develop regulations for monitoring selected unregulated contaminants (the UCMR) by August 1999 and every five years thereafter (EPA, 1998b). Unregulated contaminant monitoring is currently described under existing SDWA regulations (Title 40, Code of Federal Regulations, Part 141). However, the 1996 amendments require (1) development of a new list, which cannot exceed a total of 30 contaminants, (2) use of a representative sample of public water systems serving 10,000 or fewer people, (3) placement of data in the NCOD (when operational), and (4) consumer notification of monitoring results. Perhaps most significantly, the SDWA Amendments of 1996 require EPA to design the UCMR for use in developing future CCLs, making decisions about whether to regulate a contaminant, and promulgating subsequent regulations (EPA, 1997c). Contaminants from the CCL categorized as requiring additional occurrence data will also provide the primary

TABLE 1-1 1998 Drinking Water Contaminant Candidate List (CCL)

Microbiological Contaminants

Acanthamoeba (guidance expected for contact lens wearers)
Adenoviruses
Aeromonas hydrophila
Calciviruses
Coxsackieviruses
Cyanobacteria (blue-green algae), other freshwater algae, and their toxins
Echoviruses
Helicobacter pylori
Microsporidia (Enterocytozoon and Septata)
Mycobacterium avium intracellulare (MAC)

Chemical Contaminants	CASRN[a]
1,1,2,2-tetrachloroethane	79-34-5
1,2,4-trimethylbenzene	95-63-6
1,1-dichloroethane	75-34-3
1,1-dichloropropene	563-58-6
1,2-diphenylhydrazine	122-66-7
1,3-dichloropropane	142-28-9
1,3-dichloropropene	542-75-6
2,4,6-trichlorophenol	88-06-2
2,2-dichloropropane	594-20-7
2,4-dichlorophenol	120-83-2
2,4-dinitrophenol	51-28-5
2,4-dinitrotoluene	121-14-2
2,6-dinitrotoluene	606-20-2
2-methyl-phenol (*o*-cresol)	95-48-7
Acetochlor	34256-82-1
Alachlor ESA and other acetanilide pesticide degradation products	N/A
Aldrin	309-00-2
Aluminum	7429-90-5
Boron	7440-42-8
Bromobenzene	108-86-1
DCPA mono-acid degradate	887-54-7
DCPA di-acid degradate	2136-79-0
DDE	72-55-9
Diazinon	333-41-5
Dieldrin	60-57-1
Disulfoton	298-04-4
Diuron	330-54-1
EPTC (*s*-ethyl-dipropylthiocarbanate)	759-94-4
Fonofos	944-22-9
Hexachlorobtadiene	87-68-3
p-isopropyltoluene (*p*-cymene)	99-87-6
Linuron	330-55-2
Manganese	7439-96-5

TABLE 1-1 Continued

Chemical Contaminants

Methyl bromide	74-83-9
Methyl-*t*-butyl ether (MTBE)	1634-04-4
Metolachlor	51218-45-2
Metrobuzin	21087-64-9
Molinate	2212-67-1
Naphthalene	91-20-3
Nitrobenzene	98-95-3
Organotins	N/A
Perchlorate	N/A
Prometon	1610-18-0
RDX	121-82-4
Sodium	7440-23-5
Sulfate	14808-79-8
Terbacil	5902-51-2
Terbufos	13071-79-9
Triazines and degradation product of triazines (including but not limited to Cyanizine 21725-46-2 and atrazinedesethyl 6190-65-4)	N/A
Vanadium	7440-62-2

*a*Chemical Abstracts Registry Number

SOURCE: U.S. Environmental Protection Agency, 1998a.

source of contaminants selected for inclusion in future UCMRs. EPA is currently requesting input from the public, states, and the scientific community on options for developing the UCMR.

EPA'S CONTAMINANT IDENTIFICATION METHOD

Prior to the development of the first CCL and the direct involvement of stakeholders, states, and the NRC, EPA began work on a conceptual, risk-based approach for identifying unregulated chemical and microbiological drinking water contaminants. The identified agents were those known or anticipated to occur in public drinking water supply systems and to have the potential to affect human health (EPA, 1996b). This approach was called the Contaminant Identification Method (CIM) and was intended to identify and classify contaminants into several possible regulatory and nonregulatory categories. These categories included contaminants to be placed on the CCL (for future regulatory determinations), those requiring further toxicological research, those recommended for monitoring, those needing health advisory development or other guidance, and those for which no action is required. The CIM also was to be used periodically to re-evaluate currently regulated contaminants.

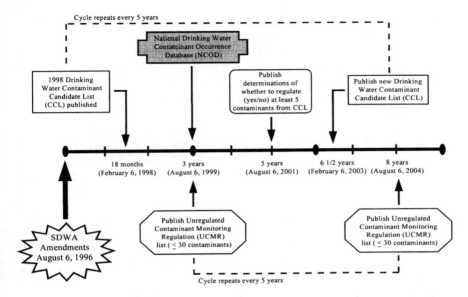

FIGURE 1-2 Time line and interaction of selected major regulatory requirements of the SDWA Amendments of 1996. SOURCE: Adapted from EPA, 1997b.

Under the CIM, the potential adverse effects, occurrence, production, use, and release of contaminants and other related factors were to be evaluated to assist in setting priorities for chemical and microbiological contaminants. The use of risk-based priorities was intended to use limited resources efficiently and to address the most important public health threats. Because of constraints associated with meeting the legislatively mandated publication deadline of February 1998 for the CCL, the CIM was not completed in time to be implemented in preparing the first CCL, and development has been postponed pending input on the CCL process from the NRC and NDWAC. Chapter 4 further discusses the CIM and the process that EPA used to prepare the draft and final 1998 CCL.

USE OF SOUND SCIENCE IN FUTURE REGULATORY DECISIONS

By congressional intent, the current and future CCLs will serve as cornerstones of EPA's future drinking water program. In making future regulatory decisions, section 1412(b)(3)(A)(i) of the amended SDWA requires EPA to use the "best available, peer-reviewed science and supporting studies conducted in accordance with sound and objective scientific practices."

It is disquieting to many nonscientists that scientific experts representing different interests can disagree markedly. There is an implicit assumption that disagreement among scientists should be rare because science is capable of objective, if not always experimental, verification. In fact, however, differences of opinion are common in science, although the arguments are spread out over many research papers and long time spans and are usually couched in careful, if not polite, language. In a regulatory proceeding, by contrast, time and space are compressed and nuances of language are erased. However, the underlying disagreements exist outside the regulatory arena as well as inside of it.

While disagreements in science are commonplace, they usually center on the applications of scientific reasoning and judgment and the conclusions drawn from such applications, not on disagreements over whether the scientist has used scientifically accepted methodology and reasoning. Epidemiological and toxicological studies—the raw material for scientific judgments of health risks caused by drinking water contamination—are like picture puzzle pieces. Depending on a particular scientist's assessment of a study's validity, the piece may be seen as clear and well defined or as fuzzy and indefinite. Depending upon that same scientist's judgment of a study's relevance, the piece may be deemed as central to the picture, a small piece on the periphery, or not part of the picture at all. The integration of (often-conflicting) epidemiological and toxicological studies in regulatory decision making is more fully discussed in Chapter 5.

The raw materials provide the puzzle pieces, but the parts do not often fit together smoothly or without gaps. Each registers different aspects of the total picture, with results that show only a portion of the whole. Placing a scientific study within a coherent picture requires the use of critical thinking, including evaluation of the part played by bias, chance, and real effect, together and separately, and judgments on what generalizations are valid. In such a complex process and with practical matters of consequence at stake, it is not surprising that differences of opinion develop and are magnified by the regulatory process. But even when so magnified, such disagreements are not artifacts of that process but are essential features of science as it is routinely practiced. The resulting disagreements are not usually evidence of flawed scientific reasoning or methodology.

This report, consequently, takes the position that scientific disagreements are the norm and do not signal a deviation from sound science. These disagreements may be based on values other than strictly scientific ones, however, this does not mean that the sides of the debate are not based on sound science. Indeed, it is not unusual for scientists to disagree on the application of sound science to public policy issues. Any scheme that affects the provision of public water is likely to engender legitimate scientific disagreement. The report also recognizes that identifying and agreeing on what is sound science is itself a difficult and error-prone enterprise. It therefore makes no recommendations on what "soundness" entails, letting the accepted mechanisms of peer regard, peer review, and scientists' habits of critical thinking continue to serve as the ultimate arbiters.

Similarly, the committee purposely declined to define what should be construed as "sufficient data," believing this is a matter of judgment that will vary with context. Any scientifically-based decision process will depend critically on the available data. It seems paradoxical that when data are sparse they are often consistent and coherent (for example, when produced by one or a few laboratories) but when data become more abundant "data gaps" appear as the problem is examined by different methods and from different perspectives. This is a natural evolution, but it makes it difficult to stipulate what should be considered "sufficient data" for a particular decision process.

THE PERSPECTIVE OF THIS REPORT

Efficient and practical provision of safe drinking water to communities of varying sizes and widely differing sources and qualities of raw water is a challenging task. Any regulation affecting this complex patchwork of local, regional, public, and private systems will almost certainly have effects beyond the intended ones. It is beyond the scope of this report to consider all the possible ramifications of its recommendations on those individuals and organizations charged with future identification and regulation of contaminants. Section 1412(b)(1)(A)(i) of the amended SDWA explicitly directs the EPA administrator to identify "contaminant[s] [that] may have an adverse effect on the health of persons." Section 1412(b)(1)(C) specifies that EPA must focus on contaminants that pose the "greatest public health concern." Therefore, in framing this report the committee has chosen to adopt an explicit public health perspective, rather than any of a number of other possible perspectives (e.g., enterprise centered, economic development, or legal). The report should be read with this qualification in mind.

REFERENCES

AWWA (American Water Works Association). 1997. Workshop Report: Mission Definition for the National Contaminant Occurrence Database (NCOD). Preliminary Draft Report. Prepared by M. M. Frey and J. S. Rosen.

Craun, G. F. 1991. Causes of waterborne outbreaks in the United States. Water Science and Technology 24:17-20.

Dzurik, A. A. 1990. Water Resources Planning. Savage, Maryland: Rowman & Littlefield, Publishers.

EPA (Environmental Protection Agency). 1994. The National Public Water System Supervision Program: FY 1993 Compliance Report. EPA/812/R/94/001. Washington, D.C.: EPA, Office of Water.

EPA. 1996a. Safe Drinking Water Act Amendments of 1996: General Guide to Provisions. EPA/810/S/96/001. Washington, D.C.: EPA, Office of Water.

EPA. 1996b. The Conceptual Approach for Contaminant Identification (Working Draft). EPA/812/D/96/001. Washington, D.C.: EPA, Office of Ground Water and Drinking Water.

EPA. 1997a. Announcement of the Draft Drinking Water Contaminant Candidate List; Notice. Federal Register 62(193): 52194-52219.

EPA. 1997b. Meeting Summary: EPA National Drinking Water Contaminant Occurrence Data Base. Washington, D.C.: RESOLVE, Inc., Contract # 68-W4-0001 prepared for EPA, Office of Ground Water and Drinking Water.

EPA. 1997c. Options for Developing the Unregulated Contaminant Monitoring Regulation: Background Document (Working Draft). EPA/815/D/97/003. Washington, D.C.: EPA, Office of Ground Water and Drinking Water.

EPA. 1998a. Announcement of the Drinking Water Contaminant Candidate List; Notice. Federal Register 63(40):10274-10287.

EPA. 1998b. Drinking Water Contaminant List. EPA/815/F/98/002. Washington, D.C.: EPA, Office of Ground Water and Drinking Water.

EPA. 1998c. Definition of a Public Water System in SDWA Section 1401(4) as Amended by the 1996 SDWA Amendments. Federal Register 63(150):41939.

Gibson, M. C., S. M. deMonsabert, and J. Orme-Zavaleta. 1997. Comparison of noncancer risk assessment approaches for use in deriving drinking water criteria. Regulatory Toxicology and Pharmacology 26:243-256.

Herwaldt, B. L., G. F. Craun, S. L. Stokes, and D. D. Juranek. 1992. Outbreaks of waterborne disease in the United States: 1989-90. Journal of the American Water Works Association 84:129-135.

Moore, A. C., B. L. Herwaldt, G. F. Craun, R. L. Calderon, A. K. Highsmith, and D. D. Juranek. 1993. Surveillance for waterborne disease outbreaks-United States, 1991-1992. Morbidity and Mortality Weekly Reporter 42:1-22.

Neal, R. A. 1985. Chemicals and safe drinking water: National and international perspective. Pp.1-8 (ch.1) in Safe Drinking Water: The Impact of Chemicals on a Limited Resource, R. G. Rice, ed. Chelsea, Mich: Lewis Publishers, Inc.

NRC (National Research Council). 1977. Drinking Water and Health. Washington, D.C.: National Academy Press.

NRC. 1997. Safe Water from Every Tap: Improving Water Service to Small Communities. Washington, D.C.: National Academy Press.

Viessman, W., and M. J. Hammer. 1985. Water management: Environmental considerations. Pp. 18-26 (ch. 2) in Water Supply and Pollution Control (fourth edition). New York: Harper & Row, Publishers.

2

Review of Existing Chemical Prioritization Schemes

A number of schemes that prioritize chemicals according to their importance as environmental contaminants have been developed by government agencies and private industries. This chapter reviews several of these existing chemical prioritization schemes. The objective is to understand the extent to which existing ranking schemes provide relevant guidance for developing a prioritization scheme for drinking water contaminants.

The committee selected a total of 10 schemes (several at the recommendation of EPA) for evaluation; these are listed in Table 2-1. The first three (the Cadmus approach, the American Water Works Association screening process, and the Regulation Development Process) are prioritization schemes specifically intended for drinking water contaminants. The next three (Waste Minimization Prioritization Tool, section 4[e] of the Toxics Substances Control Act [TSCA], and California EPA Proposition 65), are general prioritization tools for environmental contaminants. The remaining four (the Hazard Ranking System; Comprehensive Environmental Response, Compensation, and Liability Act (CERCLA) priority listing; Hazard Analysis of Releases Inventory, and the Pesticide Leaching Potential Index), are prioritization tools for specific environmental sites or media. These are included because hazardous wastes sites, contaminated sediments, and pesticide-contaminated soils all have the potential to contaminate waters that may ultimately serve as drinking water sources. The 10 schemes are intended to be representative of contaminant prioritization schemes for a variety of functions and do not include all existing contaminant ranking schemes. The prioritization schemes were evaluated using the common framework shown in Figure 2-1. "Selection of contaminant pool" refers to the universe of chemicals

TABLE 2-1   Representative Chemical Prioritization Schemes and Sources

Contaminant Prioritization Schemes Reviewed	Source[a]	Contaminant Prioritization Function
Cadmus Risk Index Approach	Cadmus Group (Cadmus Group, 1992)	Drinking water contaminants
American Water Works Association Screening Process	AWWA (RCG et al., 1993)	Drinking water contaminants
Proposed Regulation Development Process	AWWA, National Association of Water Companies, Association of Metropolitan Water Agencies, and Association of State Drinking Water Administrators (Cook, 1998)	Drinking water contaminants
Waste Minimization Prioritization Tool	EPA Office of Solid Waste and Emergency Response and Office of Pollution Prevention and Toxics (EPA, 1997)	All potential environmental contaminants
Section 4(e) of Toxics Substances Control Act	Interagency Testing Committee (Walker and Brink, 1989; Walker, 1991; Walker, 1995)	All potential environmental contaminants
State of California Safe Drinking Water and Toxic Enforcement Act of 1986 (Proposition 65)	California Environmental Protection Agency (OEHHA, 1997)	All potential environmental contaminants
Hazard Ranking System	EPA (CFR, 1997)	Hazardous waste sites
Comprehensive Environmental Response, Compensation, and Liability Act Priority List of Hazardous Substances	Agency for Toxic Substances and Disease Registry and EPA (ATSDR, 1996)	Hazardous materials
Sediment Contaminant Inventory Hazard Analysis of Releases Inventory	EPA Office of Science and Technology (EPA, 1996)	Sediment contaminants
Pesticide Leaching Potential	EPA Office of Pesticide Programs (Wolf, 1996)	Pesticides

[a] Agency, industry, or act responsible for the development of the ranking scheme.

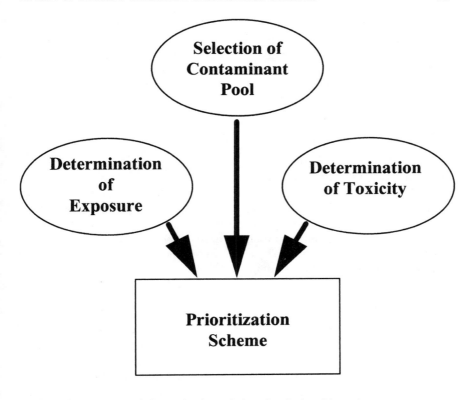

FIGURE 2-1 Framework for evaluating existing chemical ranking schemes.

considered in each prioritization scheme. "Determination of exposure" refers to the factors considered in evaluating the extent to which ecosystem and human receptors become exposed to the contaminant. "Determination of toxicity" refers to the factors considered in evaluating the potential human health effects of the contaminant. "Prioritization scheme" refers to the means by which exposure and toxicity are combined to provide a metric for ranking or prioritizing the contaminant.

REVIEW OF CHEMICAL PRIORITIZATION SCHEMES

Cadmus Risk Index Approach

The Cadmus approach is a health-risk-based methodology for ranking a candidate list of drinking water contaminants. The major components of the approach are laid out in a report entitled "Development of a Priority Pollutants

List for Drinking Water" (Cadmus Group, 1992). Using this approach, a risk index is derived to identify and prioritize chemicals that pose a potential health threat in drinking water. The risk index is based on the following chemical parameters: quantity produced, quantity released to water, persistence in water, frequency of detection in water, and toxicity to human health. The ranking scheme therefore incorporates both toxicity and exposure criteria.

Selection of Contaminant Pool

Cadmus compiled a list of candidate chemicals from a variety of sources, including the Integrated Risk Information System (IRIS) and a set of data from the Agency for Toxic Substances and Disease Registry (ATSDR) called HAZDAT. Other chemicals were obtained from the EPA's pesticides and ground water database and EPA's storage and retrieval system (STORET). Of approximately 600 chemicals, 380 were found to have defined toxicity criteria, and 155 of these were found in water. These 155 compounds formed the candidate list that was prioritized using the Cadmus approach.

Determination of Exposure

In the Cadmus approach, assessment of exposure is based on three components: annual production quantity, exposure quantity, and occurrence in water. Annual production quantity (PQ) was chosen as a possible measure of exposure because the more chemical that is produced, the greater the likelihood the chemical will be released into the environment and, ultimately, the greater the likelihood people will be exposed to it. PQ values ranges from 1 to 10, based on the division of all available (log-transformed) production quantity data into ten intervals. For example, a score of 4 indicates that the chemical had an annual production volume of 90,992 to 517,606 pounds. At the time of the development of the Cadmus risk index approach, the annual production data for all evaluated chemicals ranged from approximately 0 to 17.6 billion pounds per year.

The exposure quantity (EQ) score is a function of the quantity of a chemical released to water and the persistence of that chemical in water. The quantity released to ambient water is the sum of the quantity released to surface water as determined from the Toxics Release Inventory (TRI) database, the quantity released to surface water as determined from the Permit Compliance System chemical release database, and the quantity released to ground water as determined from the TRI database. Persistence is defined as a function of the half-life of the chemical in water and its tendency to partition to nonaqueous media as determined by the octanol-water partition coefficient. The persistence factor is assigned a value between 0.001 and 1.0. Chemicals that are most persistent in water receive a higher score than those that are degraded readily or are adsorbed onto sediments and removed from the environment. The assignment of the persistence factor value, although not clearly explained in the Cadmus report, appears in some cases to be subjective. One important simplification in assessing persistence is that chemicals introduced into the ground water are assumed to persist indefinitely. This assumption ignores the potentially important degradation or removal mechanisms of hydrolysis, adsorption, and biological degradation that can occur in an aquifer.

The occurrence in water (OW) score is a function of the frequency of detection and the maximum concentration found in ambient waters. These two noninteractive parameters are combined in an additive fashion to determine the OW score. This approach implies that a chemical may be deemed important if it is frequently detected even if at low concentrations or if it occurs at high concentrations even when found infrequently.

Determination of Toxicity

In the Cadmus approach, human health risk (HR) is defined as an average of the toxicity scores for both carcinogenic and noncarcinogenic risks. The carcinogenicity score is a weighted average of the concentration corresponding to the 10^{-5} lifetime cancer risk level; the average based upon the designations given by EPA as to whether the chemical has been found to cause cancer in laboratory animals or humans or has yet to be demonstrated as a carcinogen. The noncarcinogenicity score is also a weighted average and is based on a concentration known as the drinking water equivalent level (or DWEL) and a severity coefficient that is a measure of the type of effect produced by a particular chemical. Assignment of the weights used in the averaging relationships that ultimately determine the HR score was inadequately explained in the report and is another example of the subjective nature of the determination of the various components of the risk index.

Prioritization Scheme

The overall risk index (RI) for each chemical is computed using the following equation:

$$RI = [(W_1 * PQ) + (W_2 * EQ) + (W_3 * OW)] * W_4 HR \qquad (1)$$

where W_1 through W_4 are the weights assigned to each parameter. Within the brackets is a summation of weighted exposure information as measured by production quantity (PQ), exposure quantity (EQ), and occurrence in water (OW). This sum is then multiplied by the human health risk (HR) and by another weight factor.

An evaluation of the ranked list of chemical names at the back of the Cadmus report and the component scores attached to them illustrates an interesting point with respect to the treatment of insufficient data. If a critical element of data (e.g., defined toxicity criteria or positively detected in water) was missing for a chemical, the chemical was not included on the candidate list for subsequent risk prioritization. For example, the solvent trichloroethylene was not included on the candidate list, and yet this is the most prevalent ground water contaminant in the category of organic compounds known generally as volatile organic chemicals. It is difficult to imagine how a risk index could successfully identify chemicals that may cause problems in the future if requisite data are not available to calculate the risk index score in the first place.

American Water Works Association Screening Process

The AWWA has developed a process that evaluates a chemical's toxicity and occurrence in the environment, along with technical and economic feasibility of removing the chemical from drinking water. This process is described in a report entitled *A Screening Process for Identifying Contaminants for Potential Drinking Water Priority Listing and Regulation* (RCG et al., 1993).

Selection of Contaminant Pool

The AWWA screening process was not applied to any specific list of chemicals. Instead, the AWWA presented the methodology as part of a process that could be applied to a single contaminant or group of contaminants with similar characteristics.

Determination of Exposure

The screening process includes an evaluation of potential exposure to a chemical in the environment. A variety of data sources, including the National Organics Monitoring System, National Pesticide Survey, STORET, the Federal Reporting Data System, and U.S. Geological Survey data bases are examined for the existence of occurrence data on the chemicals of interest. Data quality is considered in the ranking or inclusion of chemicals. For example, the date of the survey and the age of the information is taken into consideration, along with published quality control and quality assurance data. Also considered are "hot spots" that may give the chemical of interest more attention than would be expected if it were simply a target of monitoring on a routine basis.

Determination of Toxicity

Data that serve as input to the toxicity screening step come from government data bases, including IRIS, Health Effects Assessment Summary Tables, Agency for Toxic Substances and Disease Registry, Hazardous Substance Data Bank,

Registry of Toxic Effects of Chemical Substances, Chemical Carcinogenesis Research Information System, and Developmental and Reproductive Toxicology Database. Each of the different data bases has different formats and methods of reporting toxicity. The practitioner must choose which of the various reporting formats should be used to categorize the chemicals.

Data quality for carcinogens is based on weight-of-evidence classification. For noncarcinogens, data quality is based on the IRIS level of confidence, which is reported as either high, medium, or low. Based on the data and an evaluation of its quality, the screening approach suggests that a compound be included on the drinking water priority list if it has at least one of the following characteristics:

1. existence—i.e., toxicity data exist in IRIS or other credible sources;
2. quality—i.e., the IRIS conclusions have high or medium confidence results and/or a carcinogen classification of A, B1, or B2; and
3. applicability—i.e., oral ingestion effects data, or high or medium confidence data on absorption after ingestion exist.

These scores are qualitative and represent an approach to assessing health effects on the basis of existence and quality of data rather than magnitude of toxic effect.

Prioritization Scheme

As described above, the AWWA screening approach first screens chemicals based on toxicity criteria. Once a chemical has been judged to have significant health effects, the screening process evaluates the potential for exposure. The AWWA screening approach also includes considerations of technical and economic feasibility. Technical feasibility is defined as the ability to control a contaminant using existing treatment technology; economic feasibility is the determination of whether reasonable treatment costs would result in an attempt to control a particular chemical.

The AWWA screening process is quite flexible and does not depend on any uniformly quantitative parameter calculation. Thus, it can include compounds on a drinking water priority list even if all the needed data to calculate a particular factor are not available. Although the toxicity to humans and prevalence of a compound in the environment should be paramount when prioritizing chemicals for regulatory action, technical and economic feasibility cannot be discounted when considering implementation and remediation strategies for those chemicals of high concern. Chemicals scoring high for toxicity and prevalence but low for technical and economic feasibility may be good candidates for new technology research.

Regulation Development Process

The Regulation Development Process (RDP) is a proposed process developed by the AWWA, National Association of Water Companies, Association of Metropolitan Water Agencies, and Association of State Drinking Water Administrators for selection of drinking water contaminants for regulation and for the analysis used to make regulatory decisions (Cook, 1998).

Selection of Contaminant Pool

The RDP has not been applied to any specific list of contaminant candidates. RDP recommends that chemicals be considered for analysis only if they have been found in ambient waters. While the RDP does not spell out what criteria should be used to select chemicals for occurrence monitoring, the process does emphasize that EPA should give significant weight to occurrence data. Thus, according to the RDP, the agency should establish a robust unregulated contaminant occurrence monitoring program that can be used to select contaminants for the Drinking Water Contaminant Candidate List (CCL). Contaminants that occur frequently with wide geographic distribution and at levels that cause health concerns, would be included on the CCL.

Determination of Exposure

The RDP assumes that the National Contaminant Occurrence Database, which is currently under development by EPA, will be the primary source of occurrence monitoring survey data on unregulated contaminants. The RDP proposes that exposure data be compiled as a plot of frequency of occurrence versus concentration. When combined with population data, this plot could be translated into a plot of national (or regional) exposure potential, providing an estimate of populations exposed to various concentrations.

Determination of Toxicity

According to the proposed scheme, toxicity data are needed to determine if there is a genuine public health threat associated with a contaminant. Health effect studies should answer the question: "Are there significant adverse health end points associated with exposure to a contaminant at concentrations seen or likely to be seen in finished drinking water?" If such data are not available, the contaminant should not be listed on the CCL.

Maximum contaminant level goals (MCLGs) may be useful screening tools, however the RDP suggests that an uncertainty range be provided with the MCLG that "describes the full range of higher MCLGs that are possible without risking disease, albeit with lower margins of safety." For nonthreshold effects (such as cancer), the RDP indicates that dose-response curves should be plotted. If extrapolation models must be used to determine risks at low doses, all uncertainties should be fully identified and carried through the analysis.

Prioritization Scheme

Under the RDP, the information from exposure analysis and health effects data would be combined to produce a frequency distribution of risk. That is, the concentration metric from the exposure analysis would be converted to a risk metric using the dose-response relationship. This conversion would translate the frequency distribution from the exposure analysis into a plot of population exposed versus risk level. Integration of the curve would yield the number of people exposed above the risk level, which could be used in evaluating the severity and magnitude of the hazard. The RDP also recommends using this plot to determine the potential for health risk reduction that would be accomplished by various maximum contaminant levels (MCLs) achievable with current treatment technologies. This estimate would serve to indicate whether setting a national MCL can produce "a meaningful reduction in the public health risk." Under the RDP, the benefits and costs of regulating a contaminant would have to be carefully considered in the risk management phase. An advantage of this approach, unlike the AWWA screening process, is that it does not eliminate contaminants from consideration based on technological or economic feasibility of treating the contaminant; instead it leaves such decisions to risk managers. A drawback to this approach, however, is that the absence of health effects data on a contaminant may significantly curtail the development of a frequency distribution of risk and would preclude inclusion of the chemical on the CCL even though there are occurrence monitoring data.

Waste Minimization Prioritization Tool

The WMPT was developed in response to the Waste Minimization National Plan by EPA's Office of Solid Waste and Office of Pollution Prevention and Toxics. It prioritizes source reduction and recycling activities based on risk (EPA, 1997). The WMPT provides a screening-level assessment of potential chronic risks to human health and the environment through prioritization of chemicals based on their persistence, bioaccumulation potential, toxicity, and quantity in the environment.

Selection of Contaminant Pool

Because the purpose of the WMPT is to assist in making decisions related to generation of environmental contaminants, the number of chemicals (4,700) covered by the WMPT is very large. The list consists primarily of chemicals on the

Toxics Substances Control Act Inventory and, in particular, those that are actually in commerce. Data exist for 880 chemicals.

Determination of Exposure

The WMPT assesses exposure based on a chemical's inherent potential to result in significant environmental exposure and thus does not include site-specific information. Exposure is based on three factors: (1) bioaccumulation potential, which is based on the octanol/water partition coefficient, bioaccumulation factor, and bioconcentration factor; (2) persistence, which is based on biodegradation rates and hydrolysis rates; and (3) potentially releasable mass of contaminant, which is based on amounts of the chemical in production waste streams. An overall measure of exposure is computed through a multiplicative relationship of factors of each of these three contributors to exposure.

Determination of Toxicity

The WMPT considers both human and ecological toxicity. Human toxicity is assessed using indicators for both cancer and noncancer effects. Indicators for cancer effects include cancer slope factors or potency factors. Indicators of a chemical's potential to cause chronic noncancer effects, such as hepatic toxicity, include EPA reference doses and reference concentrations.

Prioritization Scheme

The WMPT assigns a score to each of the three exposure factors (bioaccumulation potential, persistence, and mass) and the toxicity factor. It uses three approaches to generate these scores from quantitative data elements.

• The "binning" or "fence line" scoring approach involves comparing the quantitative value for a given chemical data element against predefined "high" and "low" threshold values, termed "fence lines." This approach is used for determining bioaccumulation potential and persistence and for some of the toxicity assessments. One advantage of the binning approach is that it accounts for chemical data that often are not very precise, and grouping data into similar "bins" avoids the false sense that such data are highly precise.
• The "continuous-scale" scoring method involves mathematically transforming the actual chemical value for a given data element into a factor score. This approach is used for determining the mass factor.
• The "decision rule" scoring method calculates factor scores based on a single or a combination of multiple data elements, following a specified set of rules. This approach is used for scoring human toxicity, depending on the data available for evaluating cancer effects.

Once factor scores are computed, they are then combined in a multiplicative relationship to obtain an overall chemical score. In recognition that the factor scores vary by orders of magnitude, the algorithm is presented as an additive relationship in which the logarithms of the factor scores are added together to generate the overall chemical score. That is,

Overall chemical score = (Human T + M + P + B)
+ (Ecological T + M + P + B)   (2)

where T denotes the logarithmic toxicity factor, M the logarithmic mass factor, P the logarithmic persistence factor, and B the logarithmic bioaccumulation potential factor. It is important to emphasize that the additive relationship results simply from a mathematical transformation of a multiplicative relationship of the factor values.

Interagency Testing Committee Approach

The Interagency Testing Committee (ITC) was established in 1976 under Section 4(e) of TSCA to screen and recommend chemicals and chemical groups for consideration by the EPA administrator for priority testing and potential rule making. It consists of representatives from 15 member and liaison agencies. The EPA administrator is required to take action on the ITC recommendations within 12 months by requiring the manufacturers of these chemicals to conduct testing or by informing the public why the testing should not be implemented. By congressional mandate, the committee must revise the Priority Testing List at least every six months.

The ITC has used three chemical selection processes to screen and identify chemicals for priority testing consideration. From 1977 to 1980, the ITC's process consisted of examining large lists of chemicals and designating chemical categories that satisfy generic definitions. From 1980 to 1989, the committee used sequential exposure and biological scoring processes followed by in-depth review. Since 1989, the committee has used computerized processes to identify chemical groups that are associated with adverse health or ecological effects or that are likely to involve occupational or environmental exposure. These computerized processes were developed to evaluate several thousand chemicals and to incorporate several feedback loops to ensure that chemicals are reconsidered as new estimates or new data become available (Walker and Brink, 1989; Walker, 1991; Walker, 1995).

Selection of Contaminant Pool

Between 1977 and 1983 the ITC conducted scoring exercises focused on high annual production chemicals (e.g., >10^6 kg/year), analogues of known carcinogens, and chemicals of concern to federal or state agencies and others. In 1986, the ITC organized expert panels to identify substructures indicative of chemicals with potential to cause adverse human health and ecological effects.

Determination of Exposure

Prior to 1986, a panel of experts assigned consensus scores to 11 exposure factors, including annual production, fraction released in the plant, number of workers potentially exposed, fraction released to the environment, number of people exposed in the general population, frequency of general population exposure, intensity of general population exposure, persistence, penetrability, influence on the environment, and bioaccumulation potential. Scores ranged from 0 to 3. Consensus scores for the 11 exposure factors were incorporated into algorithms developed to estimate occupational, general population, and environmental exposures.

Currently, ITC's computerized system assesses the potential for environmental exposure by matching Chemical Abstract Service (CAS) numbers of a large quantity of discrete organic chemicals in a computerized database with CAS-numbered chemicals in publicly available sources of exposure data (e.g., STORET database and the Chemical Evaluation Search and Retrieval System).

Determination of Toxicity

Prior to 1986, biological scoring was conducted to provide scores for mutagenicity, carcinogenicity, and developmental toxicity or reproductive effects, because TSCA section 4(e) directs the ITC to give priority attention to chemicals with the potential to cause "gene mutations, cancer and birth defects." Biological scoring was also conducted to provide scores for acute toxicity, subchronic toxicity, bioconcentration, and ecotoxicity, because the ITC considered these to be critical factors for estimating the biological effects that could be caused by chemicals. Each chemical was scored for each biological effect by a number of experts, who reviewed available data or predicted toxicity based on structure-activity relationships. Assigned scores ranged from -3 (strong suspicion of an adverse effect, but no data) to +3 (data supporting an adverse effect at the dose level). In averaging specific effect scores assigned to a chemical, no mixing of positive and negative scores was permitted. Discrepancies were discussed among the experts and resolved.

Under the current ITC approach, substructures likely to be associated with chemicals causing adverse health or ecological effects are used to probe a large universe of chemicals on the TSCA inventory. The process for selecting chemical substructures involves soliciting the opinions of internationally recognized experts with substructure questionnaires developed by ecological and health effects panels. The ITC has conducted reliability assessments of substructures used to identify potentially hazardous groups by retrieving and analyzing toxicity data on discrete chemicals in the chemical groups and supplementing these data with data extracted from studies submitted under TSCA.

Prioritization Scheme

From 1980 to 1986, the ITC used sequential exposure and biological scoring processes to prioritize chemicals for more in-depth review. By 1984, the ITC was aware that the methods used during these years had identified hundreds of easy-to-recognize chemicals in need of testing, and that it was becoming more difficult to identify priority chemicals. ITC then developed a new system, the Substruc-ture-based Computerized Chemical Selection Expert System, which used cost-effective, computerized processes that simultaneously integrate effects and expo-sure information. Chemicals from sources of environmental monitoring data and substructures with potential to cause adverse health or ecological effects are assigned codes, and these values are used to select chemicals with environmental exposure and adverse effect potential. (Chemicals with no or low effects poten-tial or low or no exposure potential are removed from the list and recycled.) This list of chemicals is then screened for low consumption chemicals. A minimum production-importation ceiling of 22,000 kg/year is used to select chemicals. An algorithm is then applied for scoring chemicals for potential adverse effects, potential exposure, and production volume. Information profiles are developed on those chemicals with high scores and are reviewed at the ITC Chemical Selection Workshop, where chemicals are selected for in-depth review based on consensus decisions. The ITC approach of combining automated sorting of chemicals with expert input is advantageous in that it allows a large number of chemicals to be considered while still allowing for expert judgment.

State of California Safe Drinking Water and Toxic Enforcement Act of 1986 (Proposition 65)

In 1986, the State of California passed the Safe Drinking Water and Toxic Enforcement Act, better known as Proposition 65 (OEHHA, 1997). This act requires the governor of California to publish lists of chemicals known to cause cancer and reproductive toxicity. The California Environmental Protection Agency's (Cal/EPA) Office of Environmental Health Hazard Assessment (OEHHA) is the lead agency responsible for implementation of Proposition 65. Two committees of the Science Advisory Board, known as the Carcinogen Iden-tification Committee and the Developmental and Reproductive Toxicant Identifi-cation Committee, serve as the state's qualified experts for rendering an opinion as to whether a chemical is known by the state to cause cancer or developmental and reproductive toxic effects.

Selection of Contaminant Pool

Under Proposition 65, OEHHA has established a tracking database to categorize chemicals, their relevant data, and evaluation status. Chemicals are included in the tracking database as a result of suggestions by state agencies and other sources, or based on literature searches. The sources of literature include scientific journals, the Chemical Carcinogenicity Research Information System, the carcinogenic potency database compiled by Dr. L. Gold and published by *Environmental Health Perspectives*, the National Cancer Institute's survey of compounds tested for carcinogenic activity, and pesticide registrant data submitted to Cal/EPA. The basis for identifying a chemical as a potential candidate may be, for example, positive cancer or reproductive toxicity bioassays or evidence of very high production or use volume accompanied by evidence of relevant toxicity. To date, more than 580 potential candidate carcinogens and more than 320 potential candidate developmental or reproductive toxicants have been entered.

Determination of Exposure

OEHHA does not use exposure information in the assignment of hazard priorities, but it accounts for this information in the selection of chemicals from the "Candidate List" for committee consideration, as discussed below. The tracking database records chemical use and occurrence information, such as whether the chemical is used in California industries, is a byproduct of industries operating in California, is a pesticide used on food crops grown or imported into California, or is a component of consumer products or drugs sold in California. Information on current restrictions on exposure to the chemical is also noted when readily available. In the absence of information specific to California, evidence of exposure, production, or use in the United States will be assumed to reflect the experience in California. A qualitative evaluation of the level of exposure concern is expressed as high, medium, low, no identified concern, or inadequate data.

Determination of Toxicity

The Proposition 65 approach assigns a draft hazard priority based on the potential for developmental or reproductive toxicity or carcinogenicity. A qualitative appraisal of this level of concern is based on an evaluation of available information, including epidemiological and animal toxicity studies and other relevant data. The list of chemicals and their draft prioritization status is then released for public and scientific comment, after which final hazard priorities are set.

Prioritization Scheme

Under Proposition 65, chemicals are selected from the tracking database for assignment of a hazard priority status based on toxicological concerns. Chemicals in the tracking database that have not yet been assigned a final priority status are included in "category I." Chemicals with a final priority status of high level of carcinogenic, reproductive, or developmental hazard concern are put on the "Candidate List." Chemicals are then chosen from the Candidate List for the preparation of a hazard identification document. The selection of chemicals from the Candidate List is based on the level of exposure concern. Thus, chemicals get on the Candidate List based solely on toxicological considerations and then are addressed in order of priority based on exposure potential.

Category II includes chemicals in the tracking database that have been assigned a final priority status other than high. This includes chemicals with a hazard priority status of medium, low, no identified concern, or inadequate data. Action is not anticipated for category II chemicals until all high priority chemicals have been identified from the tracking database, assigned to the Candidate List, and brought before the appropriate committees for evaluation. The Proposition 65 approach is thus a largely qualitative approach, involving several iterations of expert judgment.

The Hazard Ranking System

The Hazard Ranking System (HRS) is the principal mechanism that EPA uses to place hazardous waste sites on the National Priorities List of sites to be cleaned up under CERCLA. It is used to screen releases of uncontrolled hazardous substances for their potential to cause human health or environmental damage. Although it ranks sites, not chemicals, it is relevant to this evaluation because the threat posed by a contaminated site is determined to a large extent by the contaminants.

Selection of Contaminant Pool

Contaminants considered in the HRS are determined by the hazardous substances, as defined in CERCLA regulations, found at the site.

Determination of Exposure

For a given site, the HRS evaluates exposure in ground water (gw), surface water (sw), soil (s), and air (a). For each of these pathways, the system computes a "likelihood of release" factor (LR) based on the likelihood that the waste has been or will be released to each of the four pathways for transport through the environment. It computes a "waste characteristics" factor (WC) based on the toxicity, mobility, persistence, and/or bioaccumulation potential of the hazardous material. Also considered in evaluating the WC factor is the estimated quantity of the contaminant based on an assessment of the mass of waste present at the site.

The HRS also computes a target factor (T) for each pathway, considering four possible types of receptors: human individuals, human populations, natural resources, and sensitive environments. This factor is intended to describe the magnitude of the hazard with respect to the number of targets at risk. The factor is quantified by counting the number of targets involved and assessing the severity of the contamination. Severity of contamination is determined by comparing media-specific concentrations with benchmarks such as MCLGs.

Determination of Toxicity

As stated above, toxicity is a determinant used in evaluating the WC factor for a pathway. The toxicity factor for a particular hazardous substance is determined primarily from toxicological responses, represented by slope factors for cancer and reference dose values for noncancer effects. If neither of these is available, the toxicity factor can be determined from acute toxicity parameters, such as the LD_{50} (the dose of a chemical calculated to cause death in 50 percent of the test population). If no toxicity information is available for any of the hazardous substances found at the site, the toxicity factor is set at a minimum value.

Prioritization Scheme

The hazard score for each of the four pathways is computed from a multiplicative relationship of three factors: LR, WC, and T. For example, for the ground water pathway, the hazard score is computed as:

$$S_{gw} = \frac{(LR)(WC)(T)}{SF} \tag{3}$$

where SF is a scaling factor appropriate for the ground water pathway. After the hazard scores for each of the four pathways are computed, the overall site score is computed:

$$S = \sqrt{\frac{S_{gw}^2 + S_{sw}^2 + S_s^2 + S_a^2}{4}} \tag{4}$$

This relationship is an additive averaging relationship, which implies that it is not necessary for all four exposure pathways to be important for the site to rank high.

On the other hand, the overall hazard score for the site is likely to be high if all four pathways have potential to transport the contaminants.

CERCLA Priority List of Hazardous Substances

In addition to the ranking of hazardous waste sites, CERCLA also requires a ranking of the hazardous substances themselves. Section 104(I)(2) of CERCLA, as amended (42 U.S.C. 9604[I][2]) requires that the ATSDR, together with EPA, prepare a prioritized list of hazardous substances found at sites on the National Priority List (NPL). Prioritization must be based on a determination of significance of the threat to human health. To facilitate this task, ATSDR developed the HAZDAT database. This database contains information on frequency of occurrence of substances at NPL sites, potential for human exposure at these sites, and potential health effects. The prioritization scheme for the 1995 list is described in an ATSDR document published in April 1996 (ATSDR, 1996). Each substance on the CERCLA priority list is a candidate for a toxicological profile to be prepared by ATSDR and the subsequent identification of priority data needs.

Selection of Contaminant Pool

All contaminants present at NPL sites are considered for the CERCLA priority list of hazardous substances. Currently, the HAZDAT database lists more than 2,800 substances occurring at NPL sites. Only substances found at three or more NPL sites were considered for the priority list, which consists of more than 750 substances. Petroleum-related substances are excluded from the prioritization process because they are regulated by legislation other than CERCLA.

Determination of Exposure

Under this prioritization scheme, the potential for human exposure is based on two factors: relative source contribution (SC) and exposure status of populations. SC is computed as:

$$SC = \frac{\text{Theoretical Daily Dose}}{\text{RQ}} \qquad (5)$$

where RQ is the reportable quantity, which is an inverse measure of toxicity (discussed below). The theoretical daily dose is the sum of the daily doses of the contaminant from exposure to contaminated air, soil, and water. Each of these is estimated as the product of the concentration of the contaminant in that medium and the average exposure rate, using standard EPA guidelines. The concentration in each medium is representative of the maximum concentration found at a particular site and is computed as the geometric mean of maximum concentrations at observed NPL sites. The rationale for using the mean of the maximum observed concentrations, as opposed to the mean of an average of observed concentrations, is not explained in the documentation of this process.

The exposure status of populations is a categorical variable indicating whether the population has been exposed to the contaminant, has been exposed to a medium containing the contaminant, may potentially be exposed to the contaminant, or may potentially be exposed to a medium containing the contaminant. A point value is assigned to the exposure status, depending on severity, and this value is added to a logarithmic transformation of SC to give an overall score for the potential for human exposure. This approach for evaluating exposure potential is somewhat unconventional inasmuch as the score depends on a measure of toxicity (RQ).

Determination of Toxicity

The CERCLA hazardous substance prioritization scheme considers toxicity by using the RQ approach, which was developed by EPA for guidance regarding environmental releases of hazardous substances. Any person in charge of a vessel or facility from which a hazardous substance has been released in a quantity that equals or exceeds its RQ must immediately notify the appropriate authorities. RQs have been established for listed hazardous substances based on a wide variety of toxicity information, including acute toxicity, chronic toxicity, carcinogenicity, aquatic toxicity, ignitability, and reactivity. The inclusion of ignitability and reactivity is consistent with EPA's definition of hazard characteristics, even though these are generally not considered toxicity characteristics. For a substance for which an RQ value has not been established, ATSDR estimates a value for this substance and refers to it as a toxicity/environmental score (TES). An RQ (or TES) for a particular hazardous substance can be adjusted for potential hydrolysis, photolysis, or biodegradation in the environment. Thus, this parameter has information not only regarding human health effects but also regarding potential for human exposure, and yet in this prioritization scheme uses the parameter solely as a metric of toxicity.

Prioritization Scheme

In addition to measures of the potential for human exposure and toxicity, the CERCLA hazardous substance prioritization scheme uses frequency of occurrence (NPL Frequency) as a third criterion for ranking. This parameter is a scaled measure of the number of NPL sites at which a substance has been observed. The overall hazard potential of each candidate substance is computed according to the following algorithm:

$$\text{Total Score} = \frac{\text{NPL}}{\text{Frequency}} + \text{Toxicity} + \frac{\text{Potential For}}{\text{Human Exposure}} \tag{6}$$

where each quantity has been scaled to a value with a maximum of 600 points so that the total score has a possible maximum value of 1,800 points. ATSDR documentation (ATSDR, 1996) provides some rationale for the sum of the three quantities being roughly logarithmic, which would justify the additive algorithm. However, little explanation is provided for the decision to transform the toxicity metric to a suitable weighting for this criterion. Specifically, the toxicity points value is equal to 2/3 raised to the exponent of the cumulative ordinal rank, multiplied by 600, which apparently results in the desired weighting of toxicity relative to the other two components of the algorithm.

Hazard Analysis of Releases

The Hazard Analysis of Releases (HAZREL) score was developed by EPA's Office of Science and Technology as a screening-level hazard analysis to indicate sediment contamination potential and to predict where sediment problems have occurred (EPA, 1996). The objectives of the sediment inventory and analysis include generation of a relative ranking of chemicals in industrial categories using 1993 Toxics Release Inventory (TRI) and Permit Compliance System (PCS) chemical release data and prioritization of watersheds for collecting additional information to establish a baseline for future inventories. The HAZREL score is an index of the magnitude of potential sediment contamination based on specific releases, physical and chemical properties, and potential environmental risk.

Selection of Contaminant Pool

EPA selected chemicals for HAZREL ranking from the TRI and 1993 PCS chemical release data. The available list of chemicals included 25,500 individual TRI and PCS records of point source releases of 111 chemicals. About 1,020 watersheds and 31 individual industrial categories were represented by these two sources of data.

Determination of Exposure

HAZREL assesses exposure through quantification of a "fate" score ,which is calculated as the product of an air/water partitioning subfactor, a sediment adsorption subfactor, and a biodegradation subfactor.

Determination of Toxicity

HAZREL determines toxicity with respect to effects of chemicals on aquatic life. A "tox" score is calculated by taking the inverse of the sediment chemistry screening value, which is based on a combination of equilibrium partitioning and biological effects related to the protection of aquatic life. The tox score is also based on a theoretical evaluation of bioaccumulation.

Prioritization Scheme

The HAZREL score is the product of the sediment hazard score (SHS) and the annual chemical load (ACL) in pounds per year:

$$HAZREL = SHS * ACL \qquad\qquad (7)$$

The sediment hazard score is a product of fate and tox scores. Total HAZREL scores at the watershed level ranged from 0 to 312. Approximately 1,000 watersheds were evaluated, and 17 belonged in priority group 1.

The HAZREL scores are useful because they are quantitative values that can be calculated and ranked. Estimates of the fate score includes subfactors associated with physical, chemical, and biological fate, although there is little information to actually determine the subfactors necessary for an overall evaluation. However, the HAZREL score relies primarily on a determination of aquatic toxicity, which is not applicable for setting drinking water standards, and certainly the sediment aspects of the HAZREL score are not directly applicable to drinking water.

Pesticide Leaching Potential

The Groundwater Technology Section of the Environmental Fate and Groundwater Branch of EPA developed a numerical scale called the Pesticide Leaching Index, or Groundwater Leaching Index, to determine the annual risk or hazard from pesticide use with respect to ground water contamination (Wolf, 1996).

Selection of Contaminant Pool

The index has been applied only to pesticides used on apples and potatoes. There is no reason, of course, why it could not be expanded to cover other pesticides, but as yet this has not been done.

Determination of Exposure

The Pesticide Leaching Index uses a pesticide mobility index with environmental fate data and other information to compare the relative mobility of pesticides in a particular soil. Calculations for the pesticide leaching potential shown in the document describing the approach are based on Paxton sandy loam soil. The depth to ground water, a critical aspect of the pesticide leaching potential calculation, was set at 0.2 meters, which is very shallow and probably not representative of the depth of ground water in most areas of the United States.

Determination of Toxicity

No toxicity information is incorporated in the calculation of the leaching potential. Therefore, the index is only an estimate of potential exposure.

Prioritization Scheme

The Pesticide Leaching Index is a function of an attenuation factor. The index is given a score of 1, 2, or 3, depending on the level of the calculated attenuation factor. The attenuation factor is a function of soil parameters, the Henry's constant, and the pesticide half life. The index uses well developed fate and transport equations based on an understanding of how these chemicals move in soil. In general, the parameters needed for calculating the factors that make up the index are available from pesticide manufacturers who must provide these data to EPA and state agencies before a pesticide is registered and approved for use. Unfortunately, the index is narrowly focused and has only been applied to pesticides.

SUMMARY

The common theme throughout all the reviewed schemes is the prioritization of contaminants on the basis of risk to human health and/or the environment, which depends on both exposure and toxicity. The only exception to this is the Pesticide Leaching Index, which does not consider toxicity. Table 2-2 summarizes the chemical schemes with respect to the exposure and toxicity considerations incorporated into the prioritization process. Each scheme is unique in its use of data and ranking criteria, and all rely to some extent on subjective (albeit expert) judgment. This derives from the unique purpose of each prioritization scheme and the lack of a universally suitable risk ranking tool.

The exposure potential for a contaminant is determined by the likelihood of its release to the environment, the quantity released, the persistence in the environment, the proximity of the source to receptors, and the mechanisms governing its transport through the environment to a receptor. For drinking water contami-

TABLE 2-2 Summary of Chemical Contaminant Prioritization Systems

Prioritization Scheme	Exposure Determinants			Toxicity Determinants			Additional Information
	Quantity Produced/ Released	Persistence, Bioaccumulation Potential, and/ or Mobility	Quantity or Frequency of Occurrence in Environment	Human Cancer	Human Noncancer	Ecosystem Toxicity	
Cadmus	X		X	X	X		
AWWA			X	X	X		Technical and economic feasibility of control
RDP			X	X	X		Technical and economic feasibility of control
WMPT	X	X		X	X	X	
ITC	X		X	X	X	X	
CA Prop 65	X		X	X	X		
HRS		X	X	X	X	X	
CERCLA			X	X	X	X	Ignitability and reactivity
HAZREL		X					
PLP		X				X	

nants, the exposure pathway includes transport through the water distribution system. The characterization of exposure is accomplished either in an observational or a predictive fashion. Prioritization schemes that characterize exposure in an observational fashion include the CERCLA prioritization scheme, the AWWA screening process, and the proposed Regulation Development Process. These methods use monitoring data for the concentrations of contaminants in the environment to indicate exposure potential. Ideally, prioritization decisions for research and regulation would be based on environmental occurrence data, and for drinking water contaminants this may include observations of contaminants in the drinking water distribution system. Such data are often not available or are incomplete. In the absence of sufficient occurrence data, exposure may be predicted using information about the quantity of a contaminant that is produced and the frequency or rate of release to the environment, combined with estimates of persistence and mobility in the environment. The schemes that characterize exposure predictively include the WMPT, HRS, HAZREL and Pesticide Leaching Index. The ITC approach, the CADMUS approach, and California Proposition 65 employ a combined approach inasmuch as data for production and release are used along with data for observed concentrations in the environment.

The advantage of the predictive approach to evaluating exposure potential is the dependence on chemical properties that indicate fate and transport tendencies and the avoidance of site-specific environmental information. The predictive approach also relies on production and release data, and such data are easier to inventory and measure than observed concentrations in the environment, which must originate from comprehensive monitoring programs. The disadvantage, of course, is the inherent difficulty in predictive fate and transport modeling and the subsequent large uncertainty in predicted behavior. For example, none of the hazard ranking schemes evaluated in this chapter accounts for stable degradation products that may result from a variety of environmental transformations, which in some cases can be more toxic to humans than the parent compounds. Furthermore, if production rate data are used alone, they may be a poor surrogate indicator of levels of contaminant released to the environment.

The toxicological impacts of an environmental contaminant may be defined in relation to a number of receptors, broadly categorized as either human or ecological. For drinking water contaminants, prioritization according to human health impacts is relevant, whereas ecosystem impacts are not. For the prioritization schemes that account for human health impacts, toxicity is quantified using measures that indicate both cancer and noncancer effects. Typically, these are cancer slope factors and reference doses, both associated with ingestion exposure, that are taken from sources such as EPA's IRIS database. More qualitative indicators include the EPA weight-of-evidence classification scheme for carcinogenicity. The CERCLA hazardous substance prioritization scheme is unique in that it includes information about a chemical's ignitability and reactivity as part of the characterization of the toxicity score.

While all the prioritization schemes in some manner consider both exposure and toxicity, they differ in the way this information is combined. Some schemes, such as the CERCLA hazardous substance prioritization scheme, use an additive approach in which some measure of a contaminant's exposure potential is added to a measure of the contaminant's toxicity. This is equivalent to an "either/or" conceptualization in which a contaminant may rank high if it has either high exposure potential or high toxicity. The majority of schemes use a multiplicative approach, in which a measure of a contaminant's exposure potential is multiplied by a measure of the contaminant's toxicity. This is equivalent to an "and" conceptualization in which a contaminant can rank high only if it has an appreciable exposure potential and an appreciable toxicity.

If a contaminant has a high potential for exposure and is highly toxic, it will rank high under both approaches. The difference between the approaches is most apparent when considering contaminants that rank high only in one category. For example, if a contaminant is known to be very toxic, but has no known potential for exposure, then according to the multiplicative approach its rank is zero or very small. With the additive approach, such a contaminant may rank quite high. A particular problem with the additive approach is the need to consider scaling of the quantities. That is, if a toxicity metric is being added to an exposure metric, then one must decide how to scale the metrics so that they are weighted appropriately. Such decisions are often arbitrary, as in the toxicity scoring in the CERCLA hazardous substance prioritization scheme. Whether an additive or a multiplicative approach is appropriate is entirely dependent on the objectives of the prioritization scheme. It is a matter of judgment whether a contaminant should be considered important if it is only toxic, or only abundant in the environment.

Clearly, one of the greatest difficulties in constructing a prioritization scheme is determining the best way to handle uncertain and missing data. Few of the schemes are designed to address systematically the issue of statistical precision of data and how to propagate this information through the prioritization scheme. Schemes designed to serve as quantitative risk-ranking schemes rely on complete, high quality data for both exposure potential and toxicity. Of the ten schemes examined, those that fall into this category include the WMPT, the HRS, the CERCLA priority listing, the Cadmus approach, HAZREL, the Pesticide Leaching Potential, and Section 4(e) of TSCA. These schemes are extremely useful for processing large quantities of data for well-characterized contaminants, and they do not rely extensively on subjective expert judgment, which may be biased or inadequate. However, in most cases, if a contaminant is missing critical data it either is not ranked or it drops to the lowest priority. This limits the usefulness of a scheme for prioritizing emerging contaminants for which quantitative metrics of exposure and toxicity may not exist. The contaminants that may cause the biggest risk to humans may be those about which we have the least information. The AWWA screening process, the proposed Regulation Development Process, and California Proposition 65 directly consider data quality and completeness. Necessarily, these schemes rely heavily on expert judgment to make decisions in the face of uncertainty about exposure potential and health effects. These schemes do not necessarily rank contaminants, but may simply categorize them in groups with high and low priority for further action. As is clearly explained in the proposed Regulation Development Process, further action may be a decision to regulate with specification of MCLGs and MCLs, or may be a decision to prioritize research to fill data gaps.

For the present purpose of delineating an appropriate prioritization procedure for selection of contaminants currently on the CCL, none of the quantitative ranking schemes is directly applicable. This derives from the fact that contaminants may be placed on the CCL on the basis of sparse data. To make a decision in this case requires significant involvement of expert judgment. Furthermore, a variety of non-mutually-exclusive actions, including prioritization for research, promulgation of a health advisory, promulgation of a drinking water standard, or removing the contaminant from further consideration, may be recommended. A simple ranking scheme is not likely to sufficiently capture the complexity of this decision-making process.

REFERENCES

ATSDR (Agency for Toxic Substances and Disease Registry). 1996. 1995 CERCLA Priority List of Hazardous Substances That Will Be the Subject of Toxicological Profiles & Support Documents. Agency for Toxic Substances and Disease Registry, U.S. Department of Health and Human Services.

Cadmus Group. 1992. Development of a Priority Pollutants List for Drinking Water. Prepared for EPA, Office of Ground Water and Drinking Water. Washington, D.C.

CFR (Code of Federal Regulations). 1997. The Hazard Ranking System (Appendix A to Part 300). 40 CFR Ch. 1, pp.108-210.

Cook, P. 1998. A proposed regulation development process for the drinking water program: Recommendations to EPA from AWWA, NAWC, AMWA, and ASDWA. National Association of Water Companies (unpublished report).

EPA (U.S. Environmental Protection Agency). 1996. The National Sediment Contaminant Point Source Inventory: Analysis of Facility Release Data. First Draft. EPA/823/D/96/001. Washington, D.C.: EPA, Office of Science and Technology.

EPA. 1997. Waste Minimization Prioritization Tool Beta Test Version 1.0: User's Guide and System Documentation. Draft. Report No. EPA 530/R/97/019. Washington, D.C.: EPA, Office of Solid Waste and Emergency Response and Office of Pollution Prevention and Toxics.

OEHHA (Office of Environmental Health Hazard Assessment). 1997. Procedure for Prioritizing Candidate Chemicals for Consideration Under Proposition 65 by the 'State's Qualified Experts'. Sacramento, Calif.: Office of Environmental Health Hazard Assessment, California Environmental Protection Agency.

RCG et al. 1993. A Screening Process for Identifying Contaminants for Potential DWPL Listing and Regulation. Prepared for the American Water Works Association. Denver, Col.

Walker, J. D. 1991. Chemical selection by the Interagency Testing Committee: Use of computerized substructure searching to identify chemical groups for health effects, chemical fate and ecological effects testing. The Science of the Total Environment 109/110:691-700.

Walker, J. D. 1995. Estimation methods used by the TSCA interagency testing committee to prioritize chemicals for testing: exposure and biological effects scoring and structure activity relationships. Toxicology Modeling 1(2):123-141.

Walker, J. D., and R. H. Brink. 1989. New cost-effective, computerized approaches to selecting chemicals for priority testing consideration. Pp. 507-536 in Aquatic Toxicology and Environmental Fate: Eleventh Volume, G. W. Suter II and M. A. Lewis, eds. Philadelphia, Pa.: American Society for Testing Materials.

Wolf, J. 1996. Pesticide Leaching Potential to Aid in Contaminant Selection. Memorandum to E. Washington. EPA, Office of Pesticide Programs. Washington, D.C.

3

Review of Methods for
Assessing Microbial Pathogens

Historically, chemical and microbiological contaminants have been regulated in very different ways. Rather than regulating each type of microorganisms to a specific concentration, regulators have established a zero tolerance goal for microbiological contaminants and have used indicator organisms, particularly fecal coliforms, to show the possible presence of microbial contamination from human wastes. While this methodology has served well for indicating sewage contamination of surface waters and for controlling such diseases as cholera and typhoid fever, an increasing number of deficiencies with this approach have come to light in recent decades.

One deficiency in the current method used to regulate microbes is that, because of differences in survival and transport, viruses and protozoa can be present and viable in raw waters in which coliform organisms are inactive, so assessments of the safety of raw waters are sometimes too optimistic. A second problem is that some bacteria, many viruses, and many protozoa show greater resistance to many conventional treatment methods than do fecal coliforms, so assessments of the safety of treated water are sometimes too optimistic as well. A third limitation is that an increasing number of such pathogens as *Giardia* and *Legionella* are surfacing that can originate from sources other than human fecal material. Thus, the fecal indicator strategy is less relevant for these types of microorganisms.

In the past, the only database on microbiological contaminants has been a national database on waterborne disease outbreaks (discussed in Chapter 1). Until recently, this database was used to support the zero tolerance and fecal indicator regulatory strategy for microbial pathogens. More recently, EPA has begun to set goals for acceptable risk, researchers have begun to publish on methods of risk assessment for microbiological contaminants, and new techniques have evolved that suggest that the development of an occurrence database for pathogens may become possible.

There are no formal schemes, such as those reviewed in Chapter 2, for chemical contaminants that might be considered for prioritizing microbial contaminants. Nonetheless, the principles of risk assessment based on exposure potential and health impacts are similar to those for chemicals. While it is likely that some adaptation will be required, the committee believes that the time is approaching when the same risk assessment principles will be applied to the management of microbiological contaminants that are applied to chemical contaminants.

IDENTIFICATION OF MICROBIAL PATHOGENS IN DRINKING WATER

The identification of microbial hazards associated with drinking water has been accomplished in the same manner since the first documented occurrence of a waterborne disease outbreak: a cholera outbreak that was associated with contamination of the Broad Street pump in London, England, in 1855. The cause of this outbreak (contaminated drinking water) was determined through an epidemiological study. Since then, epidemiology has been the major science used to study the transmission of infectious disease through drinking water.

Epidemiology is the study of occurrence and causes of diseases in populations. The field has focused on exposure to various known or suspected toxic agents and the relationship to health outcome, using statistical methods to indicate a significant association between exposure and health. To some extent epidemiologists have attempted to describe the influence of environmental factors. The field has also been described as applying a knowledge of prevention and control to health problems; in that aspect, it is tied to risk management. Epidemiological studies are sometimes referred to as risk assessments, because there is an attempt to examine both exposure (or what is often referred to as risk factors) and health outcome in humans. However, exposure in these studies is rarely specific or quantitative for microbial contaminants, and in many cases the health hazard is defined by symptomology, rather than by the specific hazard, because a more extensive investigation is required to undertake clinical or antibody tests. Regardless, epidemiological studies can help control the occurrence of future waterborne disease outbreaks, as in the Broad Street pump study.

As discussed in Chapter 1, waterborne disease outbreaks have been investigated and reported in a national database since 1920. Thus, initial efforts to control microbial pathogens focused on bacteria. At that time, typhoid, caused by a bacterium, was the waterborne illness of most concern.

Virus outbreaks began emerging in 1950 with hepatitis A being the primary concern. From 1966 to 1970, 19 outbreaks of hepatitis A occurred. Many believe that most waterborne outbreaks caused by unidentified agents are because of viruses. Several hundred enteric viruses are possibly important agents of waterborne disease. However, information regarding the incidence of viral infections and the role of contaminated water in acquiring these is limited. Bennett et al. (1987) have reported 20 million cases of enteric viral infections and 2,010 deaths per year. Adenoviruses, which may be transmitted by the respiratory route as well, account for 10 million cases and 1,000 deaths per year, making this the most significant virus affecting U.S. populations. Rotavirus cases have been documented as the second most common viral infection and are particularly of concern for infants (MMWR, 1991).

For all virus outbreaks reported in the United States, drinking water is only one mode of transmission. Thus, epidemiological studies are needed to identify the importance of drinking water exposure. Endemic waterborne diseases of viral origin may also be important, but there is little information on the background occurrence of these diseases. Significant improvements in disease detection at endemic or low levels must occur before it is possible to assess the importance of drinking water to their occurrence.

Virus contamination of ground water is of great concern because of the resistant nature of the viral structure, which interferes with disinfection, and the colloidal size (20 nm) of viruses, which makes them easily transported through soil systems. Viruses can survive for months in ground waters (Yates and Yates, 1988; Gerba and Rose, 1989). National studies in the United States have found viruses in 20 percent to 30 percent of the ground waters. In these studies, coliforms were not predictive of viral contamination (LeChevallier, 1996). New detection techniques using the polymerase chain reaction have demonstrated that viral contamination of ground water is much more common than previously recognized (see Table 3-1).

Diarrhea has been one of the risks associated with many of the enteric viruses, such as the Norwalk virus; more serious chronic diseases have now been associated with viral infections, and these risks need to be better defined. Studies have reported, for example, that Coxsackie B virus is associated with myocarditis—the inflammation of cardiac muscular tissue (Klingel et al., 1992). This could be extremely significant given that 41 percent of all deaths in the elderly are associated with diseases of the heart. In a recent study of 43 cardiac patients, enteroviral RNA was detected in endomyocardial biopsies in 32 percent of the patients with dilated cardiomyopathy and in 33 percent of patients with clinical myocarditis (Kiode et al., 1992). In addition, there is emerging evidence that Coxsackie B virus is also associated with insulin-dependent diabetes (IDD), and this infection may contribute to a detectable increase in the number of IDD cases (Wagenknecht et al., 1991).

Protozoan diseases, specifically giardiasis, emerged as a concern in 1966. Prior to 1966, there had been only five outbreaks of amoebiasis. By the 1976-

TABLE 3-1   Virus Detection in Ground Waters in the United States

Virus	Method	Samples Positive, %
Culturable enteric viruses	Cell culture	6.8 (12/176)
Enteroviruses	PCR[a]	30
Hepatitis A virus	PCR	7
Rotavirus	PCR	13
Total viruses	PCR	39.3 (53/135)

[a]PCR is nucleic acid amplification for detection of the internal components of the virus.

1980 period, *Giardia* was the most identifiable cause of waterborne outbreaks in the U.S. Interestingly, outbreaks of *Legionella* (and several other pathogens) are not included in the waterborne disease outbreak database, although about 10,000 to 15,000 cases of Legionnaires' disease occur in the United States annually. As many as 30 percent of the respiratory diseases caused by this bacterium may be associated with tap water.

During the investigation of drinking water outbreaks, the source of the water (ground water, spring water, river water) is generally identified, along with treatment deficiency (e.g., no disinfection). More than 100 million individuals rely on ground water as a source of potable water. Only half of the community systems using ground water disinfect the water prior to distribution, while few of the noncommunity systems provide disinfection. Although ground water historically has been assumed to be safe for consumption without treatment, more than half (58 percent in 1971-94) of the reported waterborne disease outbreaks in the United States have been associated with the consumption of ground water (Craun and Calderon, 1997).

PRIORITIZATION SCHEMES FOR RULE MAKING

Selection of microbial contaminants for development of regulations has been based on reported waterborne disease outbreaks. Formal risk assessment methods utilizing occurrence databases and exposure assessment were not used until the 1980s. Haas (1983) was the first to look quantitatively at microbial risks associated with drinking water based on dose-response modeling. He examined mathematical models that could best estimate the probability of infection from the existing databases associated with human exposure experiments. Rose et al. (1991) then used an exponential model to evaluate daily and annual risks of *Giardia* infections from exposure to contaminated water after various levels of reduction through treatment. This particular study used survey data for assessing the needed treatment for polluted and pristine waters based on *Giardia* cyst occurrence. This approach was used in the development of the Surface Water Treatment Rule (SWTR) to address in particular the performance-based standards required for the control of *Giardia*; the SWTR requires a safety goal of achieving 99.9 percent reductions of *Giardia* cysts through filtration and disinfection in all surface water systems. Regulators believed this level of pathogen removal would correspond to an annual risk of no more than 1 infection per 10,000 people exposed over a year from drinking water (EPA, 1989).

Because occurrence databases were not available for enteric viruses, EPA was not able to use its goal of 1 microbial infection in 10,000 exposed persons each year to specify a treatment requirement. Instead, the SWTR mandated a treatment goal of 99.9 percent removal of viruses. This goal was derived from published information on the virus removal performance that well-operated systems with filtration and disinfection can be expected to achieve (EPA, 1987).

Since that time, more work on the occurrence of enteric viruses has been conducted. For example, it has been shown that the beta-Poisson distribution best describes the probability of infection from enteric viruses. This model has been used to estimate the risk of infection, clinical disease, and mortality for hypothetical levels of viruses in drinking water (Haas et al., 1993). Meanwhile, the development and availability of new detection techniques for viruses have allowed the creation of a meaningful occurrence database that can be used in these types of risk estimates (see Table 3-1).

Although addressed in the SWTR, there are no performance-based standards (e.g., reduction requirements) for *Legionella*, and no risk assessment was undertaken. No occurrence databases or exposure assessments exist for this bacterium.

The SWTR required that disinfection be increased to control these microorganisms, but as long as coliform standards were met, there was no way to monitor the enforcement of this rule other than requiring utilities to submit disinfectant levels and contact times. With the development of regulations to limit the levels of disinfectants and disinfectant byproducts (D/DBP), EPA recognized the possibility that efforts to reduce DBP levels could increase health risks from microbial agents. Using the Disinfection Byproducts Regulatory Analysis Model, EPA was able to examine the health and economic implications of various approaches to DBP regulation. In a direct comparison of microbial risk from *Giardia* infection to cancer risk for several DBP control scenarios, the predicted increases in *Giardia* infection were orders of magnitude higher than decreases in cancer rates. To ensure that implementation of the D/DBP rule did not increase microbial risk, the regulatory negotiating committee convened by EPA considered it necessary to review the adequacy of the existing SWTR. This revised rule, which includes regulation of *Giardia* and *Cryptosporidium*, is the Interim Enhanced Surface Water Treatment Rule (IESWTR) and is scheduled to be finalized in November 1998.

DETERMINATION OF EXPOSURE

The Information Collection Rule (ICR) is the first national program to develop occurrence data in surface waters for pathogens. Since July 1997, all utilities serving more than 100,000 people have been required to collect samples from their treatment plant influents and analyze for *Cryptosporidium, Giardia*, and enteric viruses (as well as chemical disinfection byproducts). The monitoring is scheduled to last a total of 18 months and will end in December 1998.

On February 12, 1997, EPA established the Microbial and Disinfectants/Disinfection Byproducts Advisory Committee under the Federal Advisory Committee Act (FACA) to evaluate new information and data, as well as to build consensus on the regulatory implications of new information on DBPs and pathogens that was becoming available. The advisory committee's recommendations to EPA on the proposed changes to the D/DBP rule and the IESWTR were set forth in an Agreement in Principle document dated July 15, 1997 (EPA, 1997). Because of regulatory deadlines set by Congress under the 1996 amendments to the SDWA, however, it was not possible for EPA or the FACA committee to wait for the ICR data to be collected and analyzed before changes in the IESWTR were negotiated. Proposed changes in the IESWTR were significant and include the following:

- more stringent turbidity removal requirements to control pathogens;
- establishment of a microbial benchmarking/profiling concept;
- restoration of pre-disinfection credit;
- setting of the *Cryptosporidium* MCLG at zero; and
- institution of removal requirements and credits for *Cryptosporidium*.

The results of pathogen monitoring under the ICR will be available for the next round of negotiations, scheduled to begin with a stakeholder meeting in December 1998. Negotiations for stage two of the D/DBP rule will extend over 1999 and could lead to modifications to the final SWTR, which is not expected to be completed until after the year 2000.

Analytical methods remain a critical issue for assessment of exposure to microbiological contaminants. There has been limited development and standardization of processes, however, for laboratory approval and appropriate application of both established methods (e.g., microscopy) and newer methods (e.g., immunomagnetic capture and molecular techniques). A brief summary of the historical development and contemporary use of detection and analysis methods for waterborne pathogens is included in Chapter 5. The interpretation of analytical results has also been largely neglected. Most available detection methods may be able to address some aspect of microbial occurrence (i.e., identification, quantification, viability, virulence, source, transport), but no single analytical method can be used to address all the needs of the exposure assessment (see Table 3-2).

TABLE 3-2 Examples of Exposure Factors Associated With Risks of Microbial Contaminants in Drinking Water

Exposure Factor	Data Needs
Transmission	Define fecal-oral, respiratory, contact, or multiple exposure routes.
Environmental source	Determine levels found in human waste, animal waste, sediments, biofilms, and potential loading to a water system.
Survival potential	Estimate inactivation in waste, soil, groundwater, surface water, sediments, biofilms, and determine effects of temperature, sunlight, and desiccation.
Regrowth potential	Determine growth in waste, soil, ground water, surface water, sediments, biofilms, and effects of temperature and nutrients.
Occurrence in raw water supplies	Estimate raw water type and level of contamination in different raw water types and determine spatial variations.
Resistance to treatment	Determine reduction by waste treatment, drinking water treatment, and distribution; consider resistance to disinfection, removal by filtration, etc., and adequacy of surrogates (coliform bacteria, turbidity) to evaluate removal.
Environmental transport	Quantify transport in storm events, in solids, in aerosols, to ground water, and in distribution systems.
Availability of methods[a]	Develop methods for assessing source water, identifying environmental sources, quantifying organisms, determining viability, and assessing treated water.

[a]Analytical methods must be available before other databases can be developed.
SOURCE: Adapted from Haas et al., 1998.

NEED FOR EXPOSURE AND HEALTH EFFECTS DATA

In order to examine microorganisms in drinking water and develop a prioritization scheme, data on both exposure and risk to human health are needed. Additive or multiplicative approaches could be used. However, many causes of waterborne disease are unknown; thus, the disease potential for microorganisms occurring in water needs to be examined carefully (see Table 3-3).

Outbreak investigations remain a significant component of the health effects assessment. This is the result of the extreme costs associated with outbreaks, not

TABLE 3-3   Examples of Health Factors Associated With Risks of Microbial
Contaminants in Drinking Water

Health Effects	Data Needs
Evaluation of waterborne outbreaks	Magnitude of community impact, attack rates, hospitalization and mortality, demographics, sensitive populations, level of contamination, duration, medical costs, community costs, course of immune response and secondary transmission
Evaluation of endemic disease	Incidence, prevalence, geographic distribution, temporal distribution, percentage associated with various transmission routes (i.e., water versus food), demographics, sensitive populations, hospitalization, individual medical costs, antibody prevalence, infection rates, and illness rates
Immune status	Protection of sensitive populations, lifetime protection versus temporary protection, effects of age
Description of microbial pathogens	Mechanism of pathogenicity, virulence factors, virulence genes, antibiotic resistance
Disease description	Types of disease, duration, severity, medical treatment and costs, days lost, chronic sequelae, contributing risks (i.e., pregnancy, nutritional status, lifestyle, immune status)
Methods for diagnosis[a]	Availability for routine use, special use needs, ease of use, cost, time

[a]Clinical diagnostic tests must be available before other databases can be adequately established.
SOURCE: Adapted from Haas et al., 1998.

only in medical care and days lost from work but in costs accrued in assessment of the outbreak, recall of food products, boil orders, communication efforts, remediation, and future safety efforts. The waterborne disease outbreak in Milwaukee in 1993 cost the community an estimated $25 billion, not including subsequent costs of aversion behavior because of loss of confidence in the water supply (e.g., purchase of bottled water and point-of-use devices to further treat the water). As noted in Chapter 1, investigation and reporting of waterborne disease outbreaks is not mandatory, the quality and completeness vary from state to state, and only a small proportion of the risks are identified.

Treatability of microorganisms by water processes will remain a significant part of exposure assessment. While water treatment, such as chlorination, may readily control some microbial risks, such as *Shigella* or *Campylobacter*, the reliability of treatment and the potential for growth of microbial pathogens in the water distribution system must be included in any risk assessment. Given the high risk of violations of the coliform standard (primarily in small public water systems), if disinfection failure continues to occur in a large percentage of facilities using highly polluted water supplies, the risk could become significant. It is critical that occurrence databases are developed for microorganisms that may exhibit a high level of virulence in water in order to determine the potential effects of treatment failures. Also at issue is the question of how much treatment and how much risk reduction is appropriate and acceptable. Because zero tolerance has been maintained as the goal for microbial contaminants for so long, the idea of a non-zero maximum contaminant level has not been debated nor formalized for most microbial contaminants.

SUMMARY

Historically, microbial contaminants in drinking water have not been individually prioritized for regulation. Rather, microbial contaminants have been controlled by specifying treatment methods for various types of source water and by monitoring for fecal coliform bacteria, which indicate possible presence of contamination but are not in themselves pathogenic. This system for regulating microbial contaminants has been relatively effective, but emerging new pathogens have raised concerns about whether the system is sufficient (Craun et al., 1997). Emerging waterborne pathogens of concern include protozoans (primarily *Giardia* and *Cryptosporidium*), *Legionella*, and several viruses (including various enteric viruses and adenoviruses). Limitations in data on health effects of these organisms and levels of human exposure make it difficult to establish specific priorities for their future regulation. While a tremendous amount of resources have been devoted to the control of chemical hazards in the environment (including drinking water contaminants), related expertise for controlling microbial contaminants is far less developed, even though the majority of reported waterborne illness outbreaks are known or thought to be caused by microorganisms.

REFERENCES

Bennett ,J. V., S. D. Homberg, and M. F. Rogers. 1987. Infectious and parasitic diseases. Pp. 102-114 in Closing the Gap: The Burden of Unnecessary Illness, R. W. Amber and H. B. Dull, eds. New York: Oxford University Press.

Craun, G. F., and R. Calderon. 1997. Microbial risks in groundwater systems epidemiology of waterborne outbreaks. Pp. 9-20 in Under the Microscope: Examining Microbes in Groundwater. Denver, Colo.: American Water Works Association.

Craun, G. F., P. S. Berger, and R. L. Calderon. 1997. Coliform bacteria and waterborne disease outbreaks. Journal of the American Water Works Association 89(3):96-104.

EPA (Environmental Protection Agency). 1987. National Primary Drinking Water Regulation; Filtration and Disinfection; Turbidity, *Giardia Lamblia*, Viruses *Legionella*, and Heterotrophic Bacteria; Proposed Rule. Federal Register 52(212):42194.

EPA. 1989. National Primary Drinking Water Regulations; Filtration and Disinfection; Turbidity; *Giardia lamblia*, Viruses, *Legionella*, and Heterotrophic Bacteria. Federal Register 54(124): 27486-27541.

EPA. 1997. National Primary Drinking Water Regulations: Interim Enhanced Surface Water Treatment Rule Notice of Data Availability; Proposed Rule. Federal Register 62(212):59485-59557.

Gerba, C. P., and J. B. Rose. 1989. Viruses in source and drinking water. Ch. 19 in Advances in Drinking Water Microbiology Research, G. A. McFeters, ed. Madison, Wis: Science Technology.

Haas, C. N. 1983. Estimation of risk due to low doses of microorganisms: A comparison of alternate methodologies. American Journal of Epidemiology 118:573-582.

Haas, C. N, J. B. Rose, C. Gerba, and S. Regli. 1993. Risk assessment of virus in drinking water. Risk Analysis 13:545-552.

Haas, C. N., J. B. Rose, and C. P. Gerba, eds. 1998. Quantitative Microbial Risk Assessment. New York: John Wiley and Sons.

Kiode, H., Y. Kitaura, H. Deguchi, A. Ukimura, K. Kawamura, and K. Hirai. 1992. Genomic detection of enteroviruses in the myocardium studies on animal hearts with Coxsackievirus B3 myocarditis and endomyocardial biopsies from patients with myocarditis and dilated cardiomyopathy. Japanese Circulation Journal 56:1081-1093.

Klingel, K., C. Hohenadl, A. Canu, M. Albrecht, M. Seemann, G. Mall and R. Kandolf. 1992. Ongoing enterovirus-induced myocarditis is associated with persistent heart muscle infection: Quantitative analysis of virus replication, tissue damage and inflammation. Proceedings of the National Academy of Sciences 89:314-318.

LeChevallier, M. L. 1996. What do studies of public water system groundwater sources tell us? Under the Microscope: Examining Microbes in Groundwater, Sept. 5-6. Lincoln, Nebr.: The Groundwater Foundation.

MMWR (Morbidity and Mortality Weekly Report). 1991. Rotavirus surveillance—United States, 1989-1990. Morbidity and Mortality Weekly Report 40(5):80-81, 87.

Rose, J., C. N. Haas, and S. Regli. 1991. Risk assessment and control of waterborne giardiasis. American Journal Public Health 1:709-713.

Wagenknecht, L. E., J. M. Roseman, and W. H. Herman. 1991. Increased incidence of insulin-dependent diabetes mellitus following an epidemic of Coxsackievirus B5. American Journal of Epidemiology 133:1024-1031.

Yates, M. V., and S. R. Yates. 1988. Modeling microbial fate in the subsurface environment. Critical Reviews in Environmental Control 17:307-343.

4

Approach Used to Develop the 1998 CCL

Among other changes, the Safe Drinking Water Act (SDWA) Amendments of 1996 significantly restructured the development process for drinking water standards that was established under the 1986 SDWA amendments. Prior to enactment of the 1996 amendments, every three years EPA was required to publish a drinking water priority list (DWPL) of contaminants that served as candidates for future regulation. EPA was also required to develop standards for 25 new DWPL contaminants every three years. Now, instead of publishing a DWPL and then regulating 25 new contaminants every three years, the 1996 amendments require EPA to "publish a list of contaminants, which, at the time of publication, are not subject to any proposed or promulgated national primary drinking water regulation, which are known or anticipated to occur in public water systems, and which may require regulation under this title." As discussed in Chapter 1, this new list, the Drinking Water Contaminant Candidate List (CCL), will provide the basis for deciding whether to regulate at least five new contaminants every five years. EPA published a draft of the first CCL on October 6, 1997, following a two-month public comment period, and published the final CCL on March 2, 1998 (EPA, 1998a). This chapter briefly summarizes the background and development processes for the draft and final versions of the current CCL.

EPA'S CONTAMINANT IDENTIFICATION METHOD

As discussed in Chapter 1, shortly after passage of the SDWA Amendments of 1996 EPA began work on a conceptual, risk-based approach, the Contaminant

Identification Method (CIM), to identify unregulated chemical and microbiological drinking water contaminants as priorities for its drinking water program (EPA, 1996). EPA originally developed completely separate risk-based approaches for prioritizing unregulated chemical and microbiological contaminants under the CIM.

In brief, the approach for chemical contaminants consisted of four hierarchical stages: (1) initial identification, (2) preliminary screening, (3) ranking and risk assessment, and (4) program decisions. Ultimately, EPA used only concepts from the first two stages of the chemical CIM approach to identify and select contaminants for inclusion on the CCL. No attempt was made to rank and prioritize contaminants quantitatively for future regulatory action. As noted in Chapter 1, this was largely because of time constraints associated with meeting the legislatively mandated publication deadline of February 6, 1998 (see Figure 1-1).

At the time of the CIM working draft report, EPA was uncertain whether a feasible and objective approach could be developed for microbiological contaminants using the same or similar criteria as that used for chemical contaminants. EPA considered two basic approaches for identifying and prioritizing pathogens: (1) preparation of an initial list of known and potential pathogens that would be peer reviewed and expanded by an expert panel of microbiologists and public health specialists and (2) development of a conceptual risk-based approach utilizing weighted criteria. As will be discussed later in this chapter, EPA chose the first approach in its development of the first CCL.

DRAFT DRINKING WATER CONTAMINANT CANDIDATE LIST

When published on October 6, 1997, the draft CCL included 58 unregulated[1] chemical contaminants and contaminant groups (further divided into data need categories) and 13 unregulated microbiological contaminants (EPA, 1997a). During development of the draft CCL, EPA consulted extensively with stakeholders (including water utilities, trade associations, and environmental groups), the Science Advisory Board, and the National Drinking Water Advisory Council's (NDWAC's) Working Group on Occurrence and Contaminant Selection.

NDWAC played an integral part in the development of the draft CCL. Established in 1975 under the authority of the Federal Advisory Committee Act, NDWAC provides independent advice to EPA on SDWA policies. Following enactment of the SDWA Amendments of 1996, NDWAC formed several work-

---

[1]Contaminants on the draft CCL were not subject to any proposed or promulgated national primary drinking water regulation, with the exception of nickel, aldicarb and its degradates, and sulfate, which were included because of pre-existing obligations to complete regulatory action for them (EPA, 1997a).

ing groups, including the Working Group on Occurrence and Contaminant Selection, to assist EPA in the implementation of many of its new and revised statutory requirements. This group included engineers, microbiologists, toxicologists, and public health scientists selected from federal, state, and local regulatory agencies, public and private water systems, and other organizations concerned with safe drinking water. The working group developed recommendations concerning which chemical contaminants to be included for initial consideration and criteria for EPA to use in narrowing this initial list. The recommendations were later endorsed by the full NDWAC. Also, at the recommendation of the working group, EPA sought external expertise on microbiological contaminants and convened a workshop of microbiologists and public health specialists to develop an initial list of current and emerging pathogens for possible inclusion on the draft CCL. The findings and recommendations from the workshop were fully adopted by the working group.

Identification and Selection of Microbiological Contaminants

Participants in the workshop used to develop a list of pathogens for potential inclusion on the first draft CCL included experts from academia, EPA and other federal agencies, and the water industry (EPA, 1997b). At the outset, EPA prepared and distributed a list of microbiological contaminants and criteria for selecting and prioritizing microbiological contaminants for initial consideration by workshop members. The initial list included protozoa, viruses, bacteria, and algal toxins. Inclusion was based on disease outbreak data, published literature documenting the occurrence of known or suspected waterborne pathogens, and other related information (EPA, 1997a).

Prior to reviewing EPA's proposed straw man criteria for evaluating microbiological contaminants, the participants established a set of baseline criteria related to an organism's (1) public health significance, (2) known waterborne transmission, (3) occurrence in source water, (4) effectiveness of current water treatment, and (5) adequacy of analytical methods (EPA, 1997b). All of the microorganisms included on EPA's initial list, as well as other potential microbiological contaminants that arose during deliberations, were individually evaluated against these criteria. This evaluation also assessed the basic research and data needs for each microorganism. When published, the draft CCL included every microbiological contaminant recommended by the workshop and subsequently adopted by NDWAC.

Identification and Selection of Chemical Contaminants

At the first meeting of the NDWAC working group, EPA proposed a total of 391 contaminants (including 25 microorganisms) from ten lists of potential drinking water contaminants as a reasonable starting point for developing the draft

CCL (EPA, 1997a). Most of the lists originated from a variety of EPA programs, including some from the Office of Water for use in the development of future drinking water priority lists. As briefly summarized in Table 4-1, eight lists were ultimately retained and combined to provide the working group with an initial list of 262 chemical contaminants for consideration. EPA made it clear that the number of contaminants on the draft and final CCL would have to be reduced from 262.

TABLE 4-1   Initial Chemical Lists Considered for Development of Draft CCL

List	Summary and Notes
1991 Drinking water priority list	56 total, not including disinfection byproducts for which regulations are being developed under the Disinfectants and Disinfection Byproducts Rule
Health advisories (HAs)	All contaminants with HAs or HAs under development in EPA's Health Advisory Program (108 total)
Integrated Risk Information System (IRIS)	48 contaminants adopted from IRIS based on a risk-based screen developed by EPA in anticipation of the 1994 DWPL
Contaminants identified by public water systems	List of 22 non-target contaminants identified in public water systems in anticipation of the 1994 DWPL
ATSDR list of contaminants found at Comprehensive Environmental Response, Compensation and Liability Act (CERCLA) sites	Top 50 contaminants from a 1995 CERCLA list of 275 prioritized hazardous substances
Stakeholder summary list	59 contaminants proposed as candidates by participants in a December 2-3, 1997 stakeholder meeting on EPA's CIM
Toxic Release Inventory (TRI) list	51 chemicals that met criteria for assessing the potential of a contaminant to occur in public water; derived from an original 1994 TRI list of 343 chemicals
Office of Pesticide Programs (OPP) ranking	65 pesticides and degradates taken from OPP ranking of pesticides from highest to lowest potential to leach to ground water

SOURCE: EPA, 1997a.

TABLE 4-2 Contaminants Deferred by EPA Based Solely on Suspected Endocrine Disruption

Amitrole	Parathion
Benomyl	Permethrin
Dicofol (Kelthane)	Synthetic Pyrethroids
Esfenvalerate	Transnonachlor
Ethylparathion	Tributyltin oxide
Fenvalerate	Vinclozolin
Kepone	Zineb
Mancozeb	Ziram
Metiram	Octachlorostyrene
Mirex	Polybrominated biphenyls
Nitrofen	Penta- to nonyl-phenols
Oxychlordane	

SOURCE: EPA, 1997a.

Deferred Groups of Potential Contaminants: Endocrine Disruptors

In developing the draft CCL, EPA initially prepared a list of contaminants that were suspected of having adverse effects on endocrine (hormonal) functions of humans and wildlife (EPA, 1997a). This list resulted, in part, from an interim EPA report that assessed this concern pending an extensive review by the National Research Council's Board on Environmental Studies and Toxicology, to be published in the late fall of 1998.

Under the SDWA, as amended, and the 1996 Food Quality Protection Act, EPA is required to establish a program to screen, assess, and test potential endocrine-disrupting contaminants (EPA, 1997a). In response, EPA established the Endocrine Disruptor Screening and Testing Advisory Council to advise EPA on implementing a testing and screening program. The advisory council completed its final report in August 1998 (EDSTAC, 1998). The report was reviewed jointly by EPA's Science Advisory Board and the Federal Insecticide, Fungicide, and Rodenticide Act Scientific Advisory Panel.

Pending completion of these reviews, EPA withdrew 21 contaminants (see Table 4-2) from consideration for the draft CCL based solely on the possibility of endocrine disruption (i.e., each chemical did not appear on any of the other nine initial lists of potential contaminants). However, several contaminants (all pesticides) implicated as endocrine disruptors were considered and included on the draft CCL for other reasons (e.g., dieldrin and metribuzin).

Deferred Groups of Potential Contaminants: Pesticides

During the development of the draft CCL, EPA sought assistance from EPA's Office of Pesticide Programs (OPP) in determining which pesticides should be

TABLE 4-3   Deferred Pesticides

Asulam	Halofenozide	Prometryn
Bensulfuron methyl	Halosulfuron	Propazine
Bentazon	Hexazinone	Prosulfuron
Bromacil	Imazamethabenz	Pyrithiobac-Na
Cadre	Imazapyr	Rimsulfuron
Chlorimuron ethyl	Imazaquin	Sulfentrazone
Chlorsulfuron	Imazethapyr	Sulfometuron methyl
Diazinon-oxypyrimidine	MCPA (methoxone)	Tebufenozide
Dicamba	Methsulfuron methyl	Terbufos sulfone
Ethylenethiourea	Nicosulfuron	Thiazopyr
Fenamiphos	Norflurazon	Triasulfuron
Fluometuron	Primisulfuron methyl	

SOURCE: EPA, 1997a.

priorities for the drinking water program (EPA, 1997a). In response, OPP provided a list of pesticides for consideration based on physical-chemical properties, occurrence, and extent of use. The list was ranked using a ground water risk score, which is a calculated potential for the pesticide to leach into ground water. Pesticides with a risk score of 2.0 or greater were included for initial consideration by the NDWAC working group (see Table 4-1).

During subsequent data evaluation and screening phases of the draft CCL, the working group decided to defer many pesticides from consideration when the risk score of 2.0 or greater was the only factor for inclusion on the CCL. In addition, several new pesticides for which no other data exist (besides a ground water risk score) were also deferred. In all, 35 pesticides were deferred pending further evaluation of their potential to occur at levels of health concern; these are listed in Table 4-3.

CONTAMINANT SCREENING AND EVALUATION CRITERIA

As previously noted, the NDWAC working group developed criteria for screening and evaluating potential chemical contaminants for inclusion on the draft CCL. The working group members adopted two general premises: (1) they would consider only chemical contaminants included on EPA's initial list that did not have National Primary Drinking Water Regulations, and (2) they would consider occurrence, or anticipated occurrence, first, before any evaluation of health effects information. EPA used data from the Storage and Retrieval Database, the Hazardous Substances Database, IRIS, published literature, and other regulatory agency reports, where available, to screen and evaluate potential contaminants.

Occurrence Criteria

If a specific chemical contaminant or group of contaminants met any portion of either of the following two occurrence elements, it was to be moved to the health effects phase of the evaluation:

1. Was the contaminant looked for and found in drinking water,[2] or in a major drinking water source,[3] or in ambient water at concentrations that would trigger concern about human health?[4]

2. If not looked for, was the contaminant likely to be found in water based on surrogates for occurrence, including known TRI releases[5] or high production volumes,[6] coupled with physical-chemical properties likely to result in occurrence in water supplies, or high OPP ground water risk scores.[7]

If both occurrence elements were negative, the contaminant was excluded from further evaluation and not included on the draft CCL (EPA, 1997a).

Health Criteria

Any contaminant that met the criteria for occurrence was subsequently evaluated using health effects criteria (EPA, 1997a). The major component of the health criteria evaluation was designed to determine whether there was any evidence, or suspicion, that a contaminant causes adverse human health effects. An affirmative response to any of the following criteria resulted in that contaminant's inclusion on the draft CCL:

---

[2]"Looked for and found" was defined as meaning the contaminant was identified in a drinking water survey that included a population of 100,000 or more, two or more states, ten or more small public water systems, or a data set such as EPA's Unregulated Contaminant Monitoring Database (which predates the SDWA Amendments of 1996).

[3]"Major drinking water source" was defined as a source of drinking water that served a population of 100,000 or more, or more than two states.

[4]Concentrations of concern were defined as those in ambient water samples that are within an order of magnitude of the level that is likely to cause health effects (e.g., a health advisory, drinking water equivalent level, cancer risk of 10^6) or as 50 percent of these risk levels if at least half the samples contained the contaminant at these levels.

[5]Using the Toxic Release Inventory, a release of 400,000 or more pounds of substance to surface water per year and physical-chemical properties indicated persistence and mobility.

[6]Production volume in excess of 10 billion pounds per year and physical-chemical properties indicated persistence and mobility.

[7]A high score was defined as 2.0 or greater; however, some pesticides were deferred because of lack of additional data (see Table 4-3).

1. *Is listed by California Proposition 65.* This initiative is also known as the Safe Drinking Water and Toxic Enforcement Act of 1986. It requires the Governor of California to publish a list of chemicals that are known by the State of California to cause cancer, birth defects, or other reproductive harm (CAEPA, 1997).

2. *Has an EPA health advisory.* A health advisory is a nonregulatory (i.e., nonenforceable) concentration of a drinking water contaminant at which no adverse health effects would be anticipated to occur over specific exposure durations, including a margin of safety (EPA, 1996).

3. *Is a known (based on human data) or likely (based on animal data) carcinogen according to EPA or the International Agency for Research on Cancer.*

4. *Has been linked to adverse effects in more than one human epidemiological study indicating adverse effects.*

5. *Has an oral value (reference dose) in EPA's Integrated Risk Information System.* An oral reference dose (or RfD) is an estimate of the concentration of a substance that is unlikely to cause appreciable risk of adverse health effects over a lifetime of exposure, including in sensitive subgroups (Barnes and Dourson, 1988).

6. *Is regulated in drinking water by another industrial country.*

7. *Is a member of a chemical class or family of known toxicity.*

8. *Has a structure-activity relationship that indicates toxicity.*

A negative response to every question resulted in the contaminant's exclusion from the draft CCL. EPA noted that the most useful health criteria elements were those that provided a health concentration of concern that could be compared to reported levels in water (e.g., health advisories). Conversely, a listing in California Proposition 65 or being a member of a chemical family of known toxicity was considered by the working group to be of only limited use in selecting contaminants for inclusion on the draft CCL (EPA, 1997a).

FINAL DRINKING WATER CONTAMINANT CANDIDATE LIST

The purpose of publishing the draft CCL prior to the final CCL was to seek public comment on various aspects of its development. To this end, EPA formally requested public comments on both the approach used to develop the draft CCL and on specific contaminants on the list (EPA, 1997a). EPA also sought comments on the data and research need categories contained in the draft CCL.

EPA received 71 comments on the draft CCL from many segments of the stakeholder community, including trade associations, environmental groups, industries, chemical manufacturers, state and local health and regulatory agencies, water utilities, and unaffiliated private citizens (EPA, 1998a). The majority of comments were generally supportive of the CCL development process, although many commenters advised that more robust criteria are needed for selecting contaminants for future CCLs (EPA, 1998b). In addition, many of the commenters provided suggestions, data, and information on specific contaminants they thought should be included or excluded from the final CCL. Approximately 60 issues, both contaminant-specific and related to the development of a process for preparing future CCLs, were raised by the commenters. A notebook containing all comments and EPA responses related to the draft CCL is available (EPA, 1998b). EPA considered all comments, data, and other information provided by the public in preparing the final CCL.

In response to requirements mandated by the SDWA Amendments of 1996, the final CCL was published on March 2, 1998, in the *Federal Register* (EPA, 1998c). In all, it contains 50 chemical and 10 microbiological contaminants and contaminant groups (see Table 1-1 for a complete alphabetical listing). Four microbiological and eight chemicals and chemical group contaminants were removed from the draft CCL, and one chemical (perchlorate) was added based on public comments and the continued input of the working group. Modifications to the draft CCL were also reviewed by the full NDWAC.

Expanding on EPA's original data and research needs categories for draft CCL chemical contaminants, the final CCL was divided into similar future action ("next step") categories, as listed in Table 4-4. The final CCL does not include the development of guidance as a separate future regulatory action category, as originally envisioned in the draft CCL and EPA's 1996 CIM. Rather, the development of guidance for specific contaminants has been integrated into the future action categories (e.g., sodium and *acanthamoeba*).

As noted by EPA, sufficient data are necessary to conduct analyses on extent of exposure and risk to populations via drinking water in order to determine appropriate regulatory action (EPA, 1998c). If sufficient data are not available, additional data must be obtained before any meaningful assessment can be made for a specific contaminant. At the time of the final CCL's publication, the "regulatory determination priorities" category of the CCL included those contaminants for which EPA had sufficient data to conduct exposure and risk analyses. Therefore, the five or more contaminants considered for regulation by August 2001, as required by the SDWA amendments, would likely be selected from this category. However, EPA cautioned that the future regulatory action categories of the final CCL were based on current information, and some movement between categories could be expected as additional data are obtained and evaluated.

The contaminants included in the research priorities category were those with significant data gaps in health, treatment, or analytical methods areas (EPA, 1998a). These represent EPA's priority contaminants for future data and research gathering. Similarly, the contaminants included in the occurrence priorities category have significant gaps in occurrence data. The newly revised Unregulated Contaminant Monitoring Regulation and the newly established National

TABLE 4-4 CCL Future Action Categories

Regulatory Determination Priorities	Research Priorities			Occurrence Priorities
	Health Research	Analytical Treatment Research	Methods Research	
Acanthamoeba (guidance)	*Aeromonas hydrophila*	Adenoviruses	Adenoviruses	Adenoviruses[a]
1,1,2,2-tetrachloroethane	Cyanobacteria (blue-green algae), other freshwater algae, and their toxins	*Aeromonas hydrophila* (blue-green algae), other freshwater algae, and their toxins	Cyanobacteria	*Aeromonas hydrophila*
1,1-dichloroethane	Calciviruses	Cyanobacteria (blue-green algae), other freshwater algae, and their toxins	Calciviruses	Cyanobacteria (blue-green algae), other freshwater algae, and their toxins[a]
1,2,4-trimethylbenzene	*Helicobacter pylori*	Coxsackieviruses (Information Collection Rule [ICR] data)	*Helicobacter pylori*	Calciviruses[a]
1,3-dichloropropene	Microsporidia	Calciviruses	Microsporidia	Coxsackieviruses (ICR data)
2,2-dichloropropane	*Mycobacterium avium intercellulare* (MAC)	Echo viruses (ICR data)	1,2-diphenylhydrazine Echo viruses (ICR data)	Echo viruses (ICR data)

Aldrin	1,1-dichloropropene	*Helicobacter pylori*	2,4,6-trichlorophenol	*Helicobacter pylori*[a]
Boron	1,3-dichloropropane	Microsporidia	2,4-dichlorophenol	Microsporidia[a]
Bromobenzene	Aluminum	*Mycobacterium avium intercellulare* (MAC)	2,4-dinitrophenol	1,2-diphenylhydrazine[a]
Dieldrin	DCPA mono-acid and di-acid degradates	Aluminum	2-methyl-phenol (*o-* cresol)	2,4,6-trichlorophenol[a]
Hexachlorobutadiene	Methyl bromide	MTBE	Acetochlor	2,4-dichlorophenol[a]
p-isopropyltoluene	Methyl-*t*-butyl ether (MTBE)	Perchlorate	Alachlor ESA	2,4-dinitrophenol[a]
Manganese	Perchlorate		Fonofos	2,4,-dinitrotoluene
Metolachlor	Sodium (guidance)		Perchlorate	2,6-dinitrotoluene
Metribuzin			RDX	2-methyl-phenol[a]
Naphthalene				Acetochlor[a]
Organotins				Alachlor ESA[a]

TABLE 4-4 Continued

Regulatory Determination Priorities	Research Priorities			Occurrence Priorities
	Health Research	Analytical Treatment Research	Methods Research	
Triazines and degradation products (including, but not limited to, cyanazine and atrazine-desethyl)				DCPA mono-acid and di-acid degradates
Sulfate				DDE
Vanadium				Diazinone
				Disulfoton
				Diuron
				s-ethyl-dipropylthiocarbamate (EPTC)
				Fonofos[a]
				Linuron
				Molinate
				MTBE
				Nitrobenzene
				Perchlorate[a]
				Prometon
				RDX[a]
				Terbacil
				Terbufos

[a]Suitable analytical methods must be developed prior to obtaining occurrence data.

SOURCE: EPA, 1998a.

Drinking Water Contaminant Occurrence Database, when operational, will be the primary sources for these data (see Chapter 1). In addition, as noted in Table 4-4, several contaminants included in the CCL occurrence priority category require the development of suitable analytical methods before occurrence data can be obtained.

SUMMARY

The SDWA Amendments of 1996 significantly restructured the development process for drinking water regulations established under the 1986 SDWA amendments. Shortly after passage of the 1996 amendments, EPA began work on a conceptual, risk-based approach, the CIM, for identifying and selecting unregulated chemical and microbiological drinking water contaminants as priorities for its drinking water program. However, further development of the CIM was postponed because of the tight time constraints stipulated by the SDWA amendments. In its place, a simplified and rapid process for identification and evaluation of chemical and microbial contaminants for inclusion on the CCL was adopted. For chemicals, the lead was taken by a working group of the NDWAC, using existing occurrence data as an initial screen, followed by ad hoc consideration of health effects data. A total of 50 chemical contaminants and contaminant groups were included on the final CCL. For microbial contaminants, an expert workshop was convened by EPA, and the results of its deliberations were directly adopted by the agency to select the 10 microbial agents included on the final CCL.

REFERENCES

Barnes, D. G., and M. Dourson. 1988. Reference dose (RfD): Description and use in health risk assessments. Regulatory Toxicology and Pharmacology 8:471-486.

CAEPA (California Environmental Protection Agency). 1997. Proposition 65 in plain language! *Proposition 65 Related Documents*. Online. California Environmental Protection Agency, Office of Environmental Health Hazard Assessment. Available: http://www.calepa.cahwnet. gov/oehha/docs/p65plain.htm [27 October 1997].

EDSTAC (Endocrine Disruptor Screening and Testing Advisory Committee). Final Report: Volume I. August 1998.

EPA (U.S. Environmental Protection Agency). 1996. The Conceptual Approach for Contaminant Identification (Working Draft). EPA/812/D/96/001. Washington, D.C.: EPA, Office of Ground Water and Drinking Water.

EPA. 1997a. Announcement of the Draft Drinking Water Contaminant Candidate List; Notice. Federal Register 62(193):52194-52219.

EPA. 1997b. EPA Drinking Water Microbiology and Public Health Workshop. Washington, D.C.: EPA, Office of Ground Water and Drinking Water, May 20-21, 1997.

EPA. 1998a. Announcement of the Drinking Water Contaminant Candidate List; Notice. Federal Register 63(40):10274-10287.

EPA. 1998b. U.S. EPA Response to Comment Document: Draft Drinking Water Contaminant Candidate List. Washington, D.C.: EPA, Office of Ground Water and Drinking Water.

EPA. 1998c. Definition of a Public Water System in SDWA Section 1401(4) as Amended by the 1996 SDWA Amendments. Federal Register 63(150): 41939-41946.

5

Selecting Contaminants on the CCL for Future Action:
Recommended Decision Process

The EPA faces a challenging task in determining which contaminants on the Drinking Water Contaminant Candidate List (CCL) warrant regulation. As explained in Chapter 2, existing algorithms for ranking environmental contaminants are of limited use for this purpose, because many of them were designed for priority setting, not necessarily for regulatory action, and because of data gaps and the need for policy judgments. This chapter presents a decision-making framework for selecting contaminants from a CCL for future action. It also discusses criteria for evaluating four categories of data—exposure, health effects, treatment, and analytical methods—that are needed for making this selection.

DECISION-MAKING FRAMEWORK

While a ranking algorithm may be appropriate for helping to determine contaminants to be listed on the CCL, this approach is not suitable for determining the appropriate disposition of contaminants on the CCL. Rather, the process requires considerable expert judgment to address uncertainties from the inevitable gaps in information about exposure potential and/or health effects; to evaluate, from a public health perspective, the many different effects that contaminants can cause; and to interpret available data in terms of statutory requirements. Therefore, such decisions necessarily involve subjective judgments, and the law designates EPA to make them.

For each contaminant on the CCL, there are three possible outcomes of EPA's decision process:

1. *Consider for immediate regulatory action*, as required by the Safe Drinking Water Act (SDWA) Amendments of 1996, if information is sufficient to judge that a contaminant "may adversely affect public health" and is known or is substantially likely to occur in public water systems with a frequency and at levels that pose a threat to public health.

2. *Drop from the CCL* if information is sufficient to determine that the contaminant does not pose a risk to public health in drinking water.

3. *Conduct additional research* on health effects and/or exposure if information is insufficient to determine whether the contaminant should be regulated.

These three outcomes are not mutually exclusive. For example, based on available evidence, EPA might choose to initiate regulatory action and issue a health advisory, while simultaneously pursuing research to fill information gaps that might result in subsequent further modifications of the regulatory level. The committee believes that public health will be served best by leaving EPA as much discretion as possible, within the limits of law.

Figure 5-1 shows in simplified outline a general decision process that the committee recommends for EPA use in deciding which of the above three outcomes (or combinations of outcomes) is appropriate for each contaminant on the CCL. The left side of the figure shows the suggested timing to progress through each step of the process. The framework applies to both chemical and microbiological contaminants; differences in either their characteristics or the information available about them do not justify separate decision processes.

The time line on Figure 5-1 is provided to help EPA allocate time and resources in order to meet the 1996 Safe Drinking Water Act (SDWA) Amendments' requirement to publish regulatory determinations for least five contaminants from the CCL by August 2001. The committee recognizes that almost one year of the originally allotted time (three and one-half years following publication of the first CCL) have already passed. Thus, while conveying the urgency with which EPA must act to reach the mandated regulatory decisions, the suggested time line should be of more direct use following the publication of future CCLs.

As indicated on the figure, the steps in the decision process are as follows:

1. Gather and analyze available health effects, exposure, and treatment and analytical methods data for each contaminant. This step should be initiated immediately. It is a standard task with which EPA staff are well familiar. While data on the ability of drinking water treatment technologies to remove the contaminant and analytical methods to measure the contaminant should be gathered at this stage to avoid delays in future regulatory action, these data should not be part of the decision about whether to regulate a contaminant. Any contaminant that poses a health risk in drinking water, as defined in the SDWA Amendments of 1996, should be considered for regulation. The second half of this

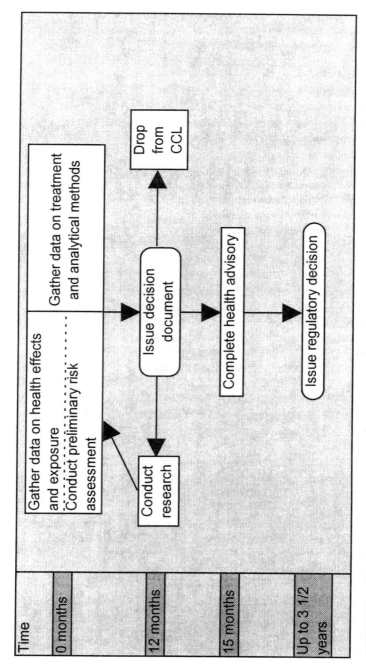

FIGURE 5-1 Phased process for setting priorities among contaminants on the CCL.

chapter describes factors to consider when gathering and assessing data on contaminant health effects, exposure, and treatment and analytical methods.

2. Conduct a preliminary risk assessment for each contaminant based on the available data. The preliminary risk assessment integrates the hazard and exposure analyses to assess the public health implications of the contaminant. It should include consideration of possible effects of the contaminant on sensitive subpopulations, such as pregnant women, infants, the elderly, and those with compromised immune systems. It should be carried out even if there are data gaps and discrepancies in order to provide a basis for an initial decision on the disposition of the contaminant and, where there are such gaps, to guide research efforts. EPA's usual approaches to risk assessment are appropriate, and the committee does not see the need to create new procedures for this step. Although a critical step in the process, the preliminary risk assessment should not be overly detailed, time consuming, or resource intensive. It should resemble risk assessments conducted by EPA under the Toxic Substances Control Act to evaluate data on new chemicals, rather than the massive multi-year risk assessments (e.g., for dioxins) that EPA often performs.

3. Issue a decision document for each contaminant describing the outcome of the preliminary risk assessment (i.e., whether the contaminant will be considered for regulation, dropped from the CCL, or retained on the CCL pending further research). This document should be issued within 12 months of compilation of the CCL. The document should describe information available to EPA at the time of the preliminary risk assessment, the weight EPA staff put on the available information and why, the reasons for EPA's decision, an action plan for implementing the decision (for example, indicating what research to conduct and how), and contacts for more information. It should be written in a language and format accessible by all interested parties.

4. Issue a health advisory for each contaminant not dropped from the CCL after the preliminary risk assessment. The health advisory should be completed within three months of the decision (within 15 months of the CCL's completion). The purpose of such an advisory should be the same as for any drinking water advisory: to alert interested parties to the possibility of a threat to public health worthy of attention and to describe the nature of the available evidence, without committing EPA to any particular future action on the contaminant. Health advisories are currently used for drinking water contaminants when the occurrence of the contaminant is not deemed widespread enough to justify imposing monitoring requirements on all utilities and to advise, even in the presence of a promulgated regulation, those to whom the regulation would not apply yet who might be vulnerable to contaminated drinking water (for example, private well owners). The committee recommends that the purpose of health advisories be expanded beyond these current uses to promulgate information about all contaminants remaining on the CCL after the preliminary risk assessment.

5. Begin compiling a regulatory package or conducting research for each contaminant remaining on the CCL after the preliminary risk assessment. This step should begin in tandem with issuance of the health advisory and should not wait until the advisory is completed. For contaminants not slated for regulation, research results should be fed back into another preliminary risk assessment, and a new decision document should be issued based on the results of this second risk assessment.

The committee's recommendation for a swift (12-month) initial decision on whether a contaminant should be put on a regulatory list is not intended to interfere with the agency's need to add and remove contaminants from such a list at any time within the five-year life of the CCL. This rapid initial action is intended to ensure that, to the extent that available information supports such an action, EPA begins as early as possible to develop a regulatory package that could support a decision to promulgate a regulation. Further, initial decisions should be made within 12 months to be sure that any information gaps (such as treatment availability and costs) standing in the way of issuing a regulation can be filled as quickly as possible.

In using this decision framework, EPA should keep in mind the importance of involving all interested parties (including regulated utilities, state and local regulators, public interest representatives, and consumers). The decision document for each contaminant should be disseminated for review by these parties, although consultation with these parties should not delay initiation of actions on the decisions that EPA has reached. Given the valid scientific disagreements noted in Chapter 1 and the way information and values are inevitably entwined, EPA would be wise to seek the insights of parties with a wide range of perspectives on contaminant priority setting during the entire decision process, not just in the period of formal regulatory procedures. Soliciting comments on the decision document will offer EPA independent perspectives and is an effective way to ensure that criteria developed after consideration of all the relevant issues have not been overlooked. In the long run, this will likely lead to a less contentious regulatory development process, if interested parties believe their views have been considered.

The EPA should also keep in mind that public health should be the guiding principle for making its decisions and that the decision to eliminate contaminants from the CCL should not be made lightly. However, there are cases when information initially used to include a contaminant on a CCL was faulty, and EPA should not be required to retain that contaminant on the list. Just as a decision to exclude a contaminant on the previous CCL from a new CCL would be explained and justified in the *Federal Register* announcing the draft CCL, a decision document would explain why EPA has decided to drop a contaminant from a CCL at other times. Conversely, if new information suggests that the contaminant is worthy of being included on the CCL after it has been eliminated from the CCL, it should be returned to the CCL. Moreover, if important new information suggests that an unregulated contaminant not listed on the CCL is being found in many water systems and may pose health risks, EPA should consider adding it to the CCL immediately, or consider invoking its immediate regulatory authority under the "urgent threats to public health" provisions of the SDWA (1412[b][1][D]).

UNCERTAINTY IN THE DECISION PROCESS

Under ideal circumstances, EPA would have a decision process that exactly selects only those contaminants whose regulation will reduce disease, disability, or death and dismisses those contaminants that have little or no effect on human health. Unfortunately, the true state of nature ("the truth") remains either unknown or shrouded in uncertainty for the majority of contaminants on the CCL. It is likely, therefore, that there will be some error in the decision process, allowing some contaminants that should be regulated to pass through while placing other, harmless contaminants on a regulatory track.

In making judgments about which contaminants to regulate, the committee recommends that EPA err on the side of public health protection. The CCL lists contaminants that are likely to pose greater risks to the public, compared to a list of randomly selected chemicals and microorganisms. For lists enriched in substances that pose risks, even a highly accurate decision process can result in many substances that need to be regulated remaining unregulated. Appendix A explains in mathematical terms why this is so. As shown in the appendix, a highly accurate decision process, when applied to such an enriched list, can still result in nearly a third of substances that need to be regulated going unregulated, while at the same time slating for regulation just three percent of substances that do not need to be regulated. Thus, for such a list, when the decision about whether or not to consider a contaminant for regulation is a close call, EPA should decide in favor of regulation.

IMPLEMENTING THE DECISION FRAMEWORK: EXAMPLES

Boxes 5-1, 5-2, 5-3, and 5-4 provide examples of how the initial data analysis step of the proposed decision framework might be implemented (or might have been implemented had the decision framework been available in the past) for four contaminants: trichloroethylene, a currently regulated contaminant; *Cryptosporidium*, which is monitored under the Information Collection Rule, but is not on the CCL; and aldicarbs and Rhodamine WT, which were both on an early draft of the CCL but were dropped before the final CCL was issued.

In presenting these examples, the committee does not seek to substitute its own judgment for EPA's. Rather, as these cases illustrate, implementing the decision framework requires a careful survey of available health effects and

BOX 5-1 Trichloroethylene:
Decision Process for a Regulated Contaminant

Trichloroethylene (TCE), a widely used organic solvent, is currently regulated in drinking water at a level of 5 micrograms per liter. The current regulation, however, was not developed as a direct result of an EPA contaminant selection and decision-making process but because Congress, in the 1986 Safe Drinking Water Act amendments, required that EPA develop a regulation for TCE and 82 other contaminants that had been slated for future regulation. At the time, Congress reacted to the belief that EPA had been too slow in developing drinking water standards following passage of the 1974 Safe Drinking Water Act and, in particular, that EPA had neglected to consider sufficiently the importance of regulating organic compounds such as TCE.

If EPA had applied the decision framework recommended in this report in 1977, the decision would likely have been (1) issue a health advisory, (2) conduct additional research on exposure and health effects, and (3) consider possible regulation, based upon the partial data that were available.

Exposure data: The primary exposure data on TCE available in 1977 were from the National Organics Monitoring Survey (NOMS), conducted in 1976-1977 (Westrick, 1990). NOMS involved the sampling of finished water (prior to distribution) from 113 water systems. The final phase of the survey found TCE above the reporting limit of 0.2-0.3 micrograms per liter in 2 of 17 ground water supplies and 17 of 88 surface water supplies, with a maximum reported concentration of 15 micrograms per liter. It is important to note that, at that time, this reported occurrence of TCE was not deemed significant. In addition, this survey was limited in that it covered mostly large water systems determined to be vulnerable to contamination. To obtain a more representative estimate of TCE occurrence EPA may have wanted to conduct additional surveys using random samples of water systems of various sizes. In fact, EPA conducted such a survey, the Community Water Supply Survey, in 1979 and found no TCE in 106 surface water systems and TCE at levels above 0.5 micrograms per liter in 14 of 330 ground water systems, with a maximum reported concentration of 210 micrograms per liter (Westrick, 1990). Thus, in 1977, additional research on exposure to TCE in drinking water likely would have been required before deciding whether to regulate TCE.

Health effects data: Health effects data on TCE were also limited in 1977. Researchers knew that TCE was metabolized to trichloroacetic acid, trichloroethanol, and small amounts of chloroform and monochloroacetate in animals (NRC, 1977), but neither the kinetics of the pathways nor any possible species differences between various strains of mice, between mice and rats, and between rats and either mice or humans was known except in the most rudimentary way. A chronic bioassay had shown liver cancer in mice but not in rats. Epidemiological data were available essentially only for high-dose occupational accident type exposures (i.e., case studies), not for the low doses found in drinking water. There was even some discussion that TCE was found as a disinfection byproduct (NRC, 1977).

Conclusions: In 1977, the existing health effects data likely would have been insufficient to drop TCE from the CCL. Therefore, TCE would warrant additional research and a health advisory. EPA would have had to decide whether or not the partial data available were sufficient to regulate the contaminant at precautionary levels.

BOX 5-2 *Cryptosporidium:*
Decision Process for an Unregulated Contaminant

Cryptosporidium, an enteric protozoan, while monitored under the Information Collection Rule, is not one of the microbial contaminants listed on the current CCL. Using the framework in this report, the preliminary risk assessment of this contaminant will likely lead EPA to a decision that it should move forward with regulatory action, although additional data (for example, on removal of this organism in different treatment processes and development of reliable monitoring methods) are likely needed to complete the regulation.

Exposure data: While other enteric protozoa have long been known to be transmitted by contaminated water, the potential for waterborne transmission of *Cryptosporidium* to humans was not recognized until the 1980s. The first documented waterborne outbreak, transmitted by well water in a small Texas community, occurred in 1984 (D'Antonio et al., 1985); a second documented outbreak occurred in Georgia in 1987 (Hayes et al., 1989). Several more outbreaks have been reported since then, with the largest occurring in Milwaukee in 1993 and affecting 400,000 individuals (MacKenzie et al., 1994; Smith and Rose, 1998). An increasing amount of research on the occurrence of *Cryptosporidium* has occurred since the first reported outbreaks. Surveys on the occurrence of oocysts were published by 1988 (Rose, 1988). Thus, the occurrence of *Cryptosporidium* in drinking water is known to be widespread enough to warrant concern.

Health effects data: Early work on *Cryptosporidium* focused on its effects on animals. First described in 1907 in the intestinal tract of mice (Tyzzer, 1907), *Cryptosporidium* was later reported to cause diarrheal disease in young mammals, particularly calves (Barker and Carbonell, 1974; Anderson and Bulgin, 1981). Mammalian isolates were shown to cause infection in other mammals, and thus this protozoan was known to cross species barriers. The first identified case in humans occurred in 1976 (Meisel et al., 1976), but cryptosporidiosis was not thought to be a cause of severe disease until the AIDS epidemic struck; the disease leads to mortality in 50 percent of cases in the immunocompromised population (MMWR, 1982). By the early 1980s, *Cryptosporidium* was known to cause illness (five to seven days of diarrhea) in populations with normal immune functions (Tzipori, 1983).

Conclusions: This organism has caused major public health concerns and is not limited to isolated water supplies. Therefore, EPA's preliminary risk assessment will likely lead EPA to decide to initiate regulatory action.

---

exposure data on the contaminant followed by policy judgments about the significance of the risk as indicated by the available data and additional research to close essential data gaps. Treatment and analytical data are not described in these examples because they are not part of the initial decision about whether a contaminant should be moved forward to the list of contaminants to consider for regulation, although assessment of these data needs to begin in tandem with exposure and health effects assessments in order to avoid delaying regulatory action and to help set research priorities.

BOX 5-3 Aldicarbs:
Decision Process for an Unregulated Contaminant

Aldicarb, a highly toxic insecticide used on such crops as potatoes, peanuts, sugar beets, soybeans, sugarcane, and cotton, is not currently regulated. The exposure and health effects data summarized below are as they existed in 1984, when aldicarb was first considered for regulation. Using historical data, one possible conclusion that EPA might have reached in 1984 if the framework had been available then is that aldicarb should be considered for a health advisory and that EPA would need to decide whether the population potentially exposed to aldicarb is sufficiently large to warrant establishing a national drinking water standard.

Exposure data (as of 1989): Aldicarb and its degradates (including aldicarb sulfoxide and aldicarb sulfone) have been found in ground water at levels that would be anticipated to be of health concern. Aldicarbs appear most frequently in agricultural areas with sandy soil, and public water supply wells in those areas are at risk of being contaminated. Water from wells near treated fields in eight states contained aldicarb at concentrations ranging from 10 to 200 micrograms per liter (EPA, 1984); these concentrations exceed health criteria suggested by the National Research Council in its 1977 report *Drinking Water and Health* (NRC, 1977). Higher levels (up to 500 micrograms per liter) have been found in New York. (EPA, 1984).

Health effects data (as of 1984): Aldicarb is known to be toxic in animals and humans by the same mechanism. Mammals readily absorb aldicarb from their gastrointestinal tract. On an acute basis, aldicarb is one of the most potent, both orally and dermally, of the widely used insecticides (rat oral LD_{50}: 0.8 mg/kg for males and 0.65 mg/kg for females; mouse oral LD_{50}: 0.3 to 0.5 mg/kg). Aldicarb is also a potent toxin in humans, as was shown by a study in groups of four adult men (NRC, 1977). At the highest dose (0.1 mg/kg), those tested experienced mild cholinergic symptoms. Cholinesterase depression occurred at lower doses (0.05 mg/kg and 0.025 mg/kg), although the findings were not statistically significant. The subchronic and chronic effects of ingesting aldicarb were studied in a 93-day rat study; two two-year rat studies; a two-year dog study; a three-generation rat study; a rat teratology study; and a mouse carcinogenicity study. These studies did not identify a more sensitive endpoint than cholinesterase inhibition. The no-observed-adverse-effect level for cholinesterase inhibition is 0.1 mg/kg/day. Based on these data, a suggested no-adverse-effect level for drinking water is 7 micrograms/liter (NRC, 1977).

Conclusions: Historical health effects data as of 1984 were sufficient to indicate that aldicarb posed a risk at concentrations found in drinking water. Therefore, according to the decision framework, EPA would have had to decide whether or not to regulate aldicarb based on its policy judgment as to whether exposure occurs with a frequency and at levels that pose a public health threat.

BOX 5-4 Rhodamine WT:
Dropping a Contaminant from the CCL

A few chemicals will "come and go" from the CCL because consensus emerges quickly that they do not present a serious threat to drinking water quality. The decision framework proposed in this report is designed to accommodate such cases by allowing a contaminant to be dropped from the CCL after release of a decision document showing that the contaminant does not pose a significant risk in drinking water.

Reasonable handling of contaminants that are judged to be of very low priority is illustrated by the case of Rhodamine WT. In the announcement of the first draft CCL (EPA, 1997), EPA included this fluorescent dye because the dye's use as a tracer in ground water flow studies apparently had resulted in detectable concentrations above the National Sanitation Foundation's (NSF's) standard of 0.1 mg/L. However, commenters on the draft list pointed out that the 0.1 mg/L standard was for drinking water and that the data that raised EPA's concern came from "ground water not associated with drinking water production," for which the NSF standard is 100 mg/L. In light of this clarification and because (1) there are no data indicating adverse health effects of Rhodamine WT and (2) the dye is used for very specific and limited purposes, EPA chose not to list Rhodamine WT on the final CCL.

Conclusions: If these data had come to light after Rhodamine WT was included on a CCL, a decision document explaining EPA's reasoning would have allowed the contaminant to be dropped from the CCL.

As noted by EPA, sufficient data are necessary to conduct analyses on extent of exposure and risk to populations via drinking water in order to determine appropriate regulatory action (EPA, 1998). If sufficient data are not available, additional data must be obtained before any meaningful assessment can be made for a specific contaminant. At the time of the final CCL's publication, the "regulatory determination priorities" category of the CCL included those contaminants for which EPA had sufficient data to conduct exposure and risk analyses. Therefore, the five or more contaminants considered for regulation by August 2001, as required by the SDWA amendments, would likely be selected from this category. However, EPA cautioned that the future regulatory action categories of the final CCL were based on current information, and some movement between categories could be expected as additional data are obtained and evaluated.

GENERAL GUIDELINES FOR EVALUATING
CONTAMINANT-RELATED DATA

Because of the variability in the types and quality of data available on different contaminants, defining precise criteria for placing contaminants in the three decision categories (regulate, drop from CCL, or research) is not possible, as the examples presented in Boxes 5-1, 5-2, 5-3, and 5-4 illustrate. Nevertheless, establishing general guidelines is possible. Below, the committee recommends such guidelines for evaluating data on contaminant exposure, health effects, and treatment and analytical methods.

Assessing Exposure Data

Exposure data should be gathered from sources that will predict the dose of drinking water contaminants for individuals, whether it be through ingestion, inhalation or dermal absorption. Table 5-1 represents a hierarchy of data types for the assessment of exposure.

Ideally, the best estimate of an individual's exposure to drinking water contaminants would be determined from samples collected at the person's tap. Such samples reflect all of the changes that might occur in the distribution system, treatment plant, and source waters that precede it. By integrating the results of a tap-sampling program, it is possible to obtain a picture of population exposure to the contaminant of interest.

Rarely is a census of tap water quality available, however. Tap sampling information is more difficult to obtain because of potential problems with access and costs. It is also prohibitively expensive to determine the tap water quality of every customer. While some utilities use consumers' taps as sample points, utilities are converting to dedicated sampling stations located on distribution system mains to obtain representative samples of the water under their control.

The second most useful sampling locations to estimate contaminant exposure are in a drinking water distribution system. Distribution sampling locations must be carefully selected to represent the characteristics of the contaminants being monitored. For example, concentrations of trihalomethanes and a variety of other disinfection byproducts change during transport through distribution systems as a result of continued exposure to chlorine. Thus, the trihalomethane regulation requires that these compounds be sampled at three average and one

TABLE 5-1 Hierarchy of Data Needed for Exposure Assessment

Concentration at the tap
Concentration in the distribution system
Concentration in finished water of the water treatment plant
Concentration in raw (source) water
Concentration in watersheds and aquifers
Concentration in historical contaminant release data
Concentrations in production data
Concentrations in biota and human tissue

maximum detention time location for each treatment facility (EPA, 1979). Sampling of distribution systems, if properly designed, can be far less costly than sampling individual taps, but it provides less precise exposure information.

If distribution system water quality information is not available, samples collected from treatment plant-finished waters can be used to represent how consumers are typically exposed to contaminants. Collecting and analyzing samples from finished water locations is especially useful if no changes in contaminant concentration or composition are expected during transport through the distribution system (i.e., for "conservative" contaminants). However, transport through the distribution system generally changes the concentration and characteristics of most contaminants. For example, the concentrations of microorganisms change between the treatment plant and the consumer's tap, because of continued action of the disinfectant or, where disinfection is inadequate, microbial regrowth in the distribution system. However, for substances (such as radon) that distribution transport might not affect, finished water sampling may be sufficient.

Similarly, determining exposure of consumers to contaminants by using data collected in a watershed for a surface water resource or samples collected from a groundwater aquifer has the potential for producing a misleading picture; many changes in contaminant concentration can occur during transport from the source to the treatment plant intake and during subsequent treatment. However, knowledge that a particular raw water source is or is not heavily polluted and the source(s) of the contaminants is always helpful.

Chemical release data or concentrations of microorganisms in discharges are examples of data that can be used for estimating how significant a contaminant in water sources could be. However, use of historical contaminant release data (or, even more removed, production data) to predict human exposure is problematic. As previously noted, a chemical's concentration and characteristics may change dramatically following production, release to the environment, and subsequent intake by humans from contaminated drinking water. Thus, only gross relationships between contaminants of vastly different release or production amounts may be possible, and even these may be misleading.

Lastly, data showing contaminant concentrations in human tissue or in plant or animal materials (i.e., biomarkers of exposure) are uncommon. This type of monitoring is expensive, is currently of unknown utility, and generally focuses on chemicals of known toxicity. Contaminant concentrations in tissues or biota are currently not a likely source for determining possible exposure to unknown chemicals, but the availability and utility of such data may increase in the future.

Criteria for Exposure Data Used in Risk Assessments

The available exposure data for a given contaminant may not be sufficient to support a defensible risk assessment. The EPA will need to determine specific, contaminant-dependent criteria for which data are acceptable for this purpose. In general, exposure data for risk assessments should have sufficient spatial and temporal coverage, and exposure should be to a minimally defined number of people. As discussed in Chapter 1, EPA needs to define terms such as "sufficient" and "minimal number of people."

The quantity and quality of monitoring data for any one contaminant will depend on the contaminant's regulatory status and the primary purpose for which the analytical work was undertaken. Four cases can be distinguished: (1) monitoring of regulated compounds required under EPA's Information Collection Rule (ICR), (2) surveys of unregulated but targeted compounds required under the Unregulated Contaminant Monitoring Rule (UCMR), (3) information to be contained in the proposed National Drinking Water Contaminant Occurrence Database (NCOD), and (4) ad hoc studies focused on particular contaminants or surveys of particular families of compounds.

As discussed in Chapter 2, the ICR requires large public water systems to monitor for microbial contaminants and disinfection byproducts. In order to help ensure that monitoring data meet specific accuracy and precision requirements, EPA established a national laboratory approval process to identify laboratories qualified to perform analyses for the ICR. In general, occurrence data acquired by water utilities under the ICR should be adequate for an exposure assessment to evaluate compounds on the CCL. Occurrence data collected under the revised UCMR and stored in the NCOD should also be adequate to identify compounds from the CCL that may require regulation. Chapter 1 reviews the regulatory development, time line, and intended use of both the UCMR and the NCOD. However, additional occurrence data (presumably from detailed ad hoc studies of particular contaminants) may be required for compounds considered priority candidates for regulation. Raw (source) water data from federal surface water monitoring programs such as the National Water Quality Assessment program (run by the U.S. Geological Survey), the National Stream Quality Accounting Network (U.S. Geological Survey), and the Environmental Monitoring and Assessment Program (EPA) should also be of acceptable quality.

Considerations for Research and Monitoring

For contaminants that do not have sufficient exposure information to conduct a preliminary risk assessment, additional research and/or monitoring will be needed. To achieve this, sensitive analytical methodologies with sufficient spatial and temporal measurements are needed for each contaminant. The first step is to develop an analytical method if one does not currently exist. This method should be precise and accurate for the given contaminant. The greatest analytical challenges lie in the identification of new contaminants and the quantification of emerging contaminants that are intrinsically difficult to measure.

As previously noted, occurrence data are a high priority, but they require considerable time and effort to collect. While designing and implementing such a monitoring program, the committee recommends that exposure concentrations be estimated from available data using models. A combination of models could be used to predict tap water concentrations of a contaminant from finished water data, environmental measurements, measurement of surrogates that are readily analyzed, or production/release data. For example, environmental measurement data for microbial contaminants could be used to model exposure from tap water using a fate-and-transport model coupled with a distribution system model. Analogous models could be used to translate production/release data for chemical contaminants into exposure concentrations.

The appropriate level of complexity of fate and transport modeling for this purpose depends on the spatial distribution of the input data. For very localized contaminant sources it may be appropriate to use a detailed, site-specific fate and transport model. For example, this type of modeling has been used to describe the migration of *Cryptosporidium* from known agricultural sources to the raw water intake in Milwaukee, Wisconsin. For widely distributed environmental contaminants, such as nonpoint-source pollutants, very simple fate-and-transport models may be used that include only parameters describing persistence and mobility in the environment. To estimate these properties, physical/chemical parameters such as the Henry's Law constant, octanol/water partition coefficient, aqueous solubility, and degradation rates are needed.

For contaminants with an existing, acceptable analytical method, monitoring data should also be collected if they are not already available. While concentrations at the tap are the ultimate goal, it would be most effective to design a monitoring plan that measures contaminants of interest in the raw water, finished water, or distribution system (depending on the best available analytical method) and then verifies exposure with selected tap monitoring. Such data are also useful for verifying the models described above. Finally, any data collected should provide representative spatial and temporal coverage needed for a defensible risk assessment.

Assessing Health Effects Data

The health effects-related information for the priority-setting process include (1) toxicological laboratory studies and data bases; (2) epidemiological studies, clinical studies, and case reports; and (3) predictive biological activity or effects models, commonly referred to as structure activity relationship (SAR) and/or quantitative structure activity relationship (QSAR) models.

The committee recommends the following general principles for assessing health effects-related criteria:

- Positive epidemiological studies should be considered of highest value for priority setting purposes even in the presence of negative toxicological studies.

• Although positive toxicological studies will take priority for regulation in most cases, negative or inconclusive epidemiological studies should be considered and an attempt should be made to explain their results when determining priority.

Human data offer several advantages over data from animal studies: (1) elimination of the uncertainty resulting from interspecies extrapolation; (2) reduction of the uncertainty caused by high-to-low-dose extrapolation, since, for example, the range between occupational exposures and likely environmental exposures is smaller than between the doses administered in animal studies and likely environmental exposures; (3) more accurate reflection of the relevant real-life exposure scenario; and (4) evaluation of the effects of the chemical on susceptible subgroups (Hertz-Picciotto, 1995; Federal Focus, 1996). However, primarily because of the bias introduced by exposure misclassification, as well as other biases, environmental and occupational epidemiologic studies may easily underestimate or miss a true adverse health effect. Therefore, it is important to evaluate all the evidence available, including animal studies and case reports, as well as epidemiologic studies (Shepard, 1994; Wartenberg and Simon, 1995).

The remainder of this section describes in more detail the nature of the information included in each health effects-related information category, the strengths and limitations of the type of information in each category, and guiding principles for using the information to evaluate CCL contaminants for further regulatory actions.

Toxicological Data

Information gained from studies in laboratory animals is commonly employed in estimating whether there might be potentially adverse human health risks associated with exposure to contaminants in drinking water. In preparing preliminary risk assessments of contaminants on the CCL, EPA should summarize in narrative form the available toxicological studies and highlight aspects relevant to the health effects the contaminant may cause.

In preparing the summaries, EPA should keep in mind that doses used to examine the toxicology of a chemical or mixture of chemicals are initially given at high levels to laboratory animals so that adverse effects can be observed, but that these high-dose studies may have limited relevance to drinking water. The primary goal of these high-dose experiments is to observe the qualitative nature of the toxicity, which includes organs and tissues involved, species differences, gender differences, time of onset, and permanence of the effects. High doses are also needed when the event of concern (for example, tumor formation) needs to be detected at a rate that would make the use of lower doses infeasible because of the large number of animals needed. Although studies of chemical toxicity usually begin with high doses, exposure to contaminants in drinking water typically results in low and continued daily doses of substances. If these doses are not completely eliminated from the body on a daily basis, they may accumulate to levels that exceed a threshold for producing toxicity or that increase substantially the risk of contracting a disease. The situation in which acute toxicity results from a large dose of a drinking water contaminant is extremely rare. Chronic toxicity resulting from lower exposures presents a more likely scenario. Chronic toxicity tests also permit time for adaptive processes (e.g., induction of metabolizing enzymes) to affect the animal's response to the chemical. These adaptive processes may exacerbate or diminish the observed toxic response. Long-term exposures also allow for longer-latency diseases to develop (e.g., cancer) in the experimental animals. Thus, greater weight should be given to toxicity data obtained from laboratory animals given lower-dose, continual exposures than to acute toxicity tests using high doses.

The toxicity measurements made in laboratory animals should be as extensive as is practical and include lethality, organ damage, tissue and cell abnormalities at the microscopic level, and relevant biochemical parameters associated with physiological dysfunction in the animal. If possible and appropriate, it is also desirable to identify doses that produce no observable effects (the no observed effect level or NOEL and/or the no observed adverse effect level or NOAEL) and doses that produce changes that represent the first evidence of overt toxicity (the lowest observed adverse effect level or LOAEL). For some outcomes, such as cancer and reproductive effects, this might not be possible, because the outcome is of an "all-or-none" variety that occurs with a low probability that would still be of importance when large populations are exposed. The sensitivity with which a toxic effect is detected may be enhanced by using laboratory animal species that have high susceptibility to the toxic agent and by using measurements that detect a nontoxic physiological or biochemical change that represents a prelude to the toxic event (i.e., a biomarker of the early effect).

For interpretation of results obtained from chronic toxicity tests, it is useful to know the blood concentration of the chemical and its toxic metabolites in the animals several times during the tests. This information helps the decision maker judge the validity of the extrapolation to other animals species and to humans. In addition, blood concentrations often produce information to estimate whether species differences in toxicity are a result of toxicokinetic (absorption, distribution, metabolism and excretion effects of the organism on the chemical) or toxicodynamic (effects of the chemical on the organism) dissimilarities among species.

When evaluating the merits of different toxicological studies, in vivo studies with relevant endpoints and a range of dose-response data should be given greater weight than in vitro studies (EDSTAC, 1998). Further, studies that show a correlation between dose and effects, that have followed good laboratory practices, and that have been peer reviewed should be given greater weight than those that do not meet those criteria.

Knowledge of the biochemical pathways through which chemicals produce deleterious effects in laboratory animals can be used to improve the accuracy and validity of the prediction of human risk. Species differences in the qualitative and quantitative aspects of chemical-induced toxicity make extrapolation between species difficult. Predictions of chemical toxicity in humans from information obtained in laboratory animals could be greatly enhanced in the future by emerging knowledge of the human genome, as well as the genome of common laboratory animal species. At the moment, the use of mechanistic information remains limited because of very substantial data gaps and inconsistencies concerning the actual mechanisms at work in humans exposed under natural conditions and the extent of variability among individuals. The EPA should consider including studies of methods for incorporating mechanistic information into assessments of health risks from contaminants in drinking water and other exposure vectors as part of its research strategy for some contaminants on the CCL. Such studies will need to consider the possibility that a single contaminant may affect health through more than one mechanistic pathway and that interactive effects may occur when multiple contaminants are present.

In the absence of information from human epidemiology, data from toxicity experiments in several laboratory animal species is usually necessary (although mechanistic information that validates the use of a particular laboratory animal species as a model for the human may obviate the need for data in several species). The availability of information from well conducted human studies that indicate a sufficiently strong association between chemical exposure through drinking water and adverse health outcomes would require fewer (or no) supporting data from animal studies. A documented biological rationale (based on results from animal studies and other relevant information) for an association between human exposure to a drinking water contaminant and a particular adverse health effect enhances a conclusion of causality in an epidemiological study. However, a lack of supportive animal data for an association between contaminant exposure and a health outcome may indicate (among other possible explanations) that a causative association may not be present or that the particular animal models used were not appropriate (e.g., arsenic).

Given a lack of sensitivity for detection of health outcomes using epidemiology, or a lack of data because the problem may not have been studied or cannot be studied in human populations, animal toxicology data must still be used to provide a human risk evaluation. Only infrequently is it found that an agent known to produce human toxicity will not produce a similar effect in some laboratory animal species when given sufficiently high doses. While all possible scenarios describing the interaction of data derived from humans and from laboratory animals have not been addressed here, it should be apparent that the application of both types of data represents the best approach to assessing the potential health risks from exposure to chemicals in drinking water.

Epidemiological Data

As for toxicological studies, in the evaluation of human data, EPA should systematically review each study or case report and summarize it in a narrative. In particular, aspects of each study/case report that might be relevant to the determination of heterogeneity of findings among the available studies/case reports should be highlighted in the narrative summaries. Such sources of heterogeneity (i.e., differences in study findings that cannot be accounted for by sampling variation) include differences in study design, in the distributions of susceptible subgroups in the study populations, and in the ability to adjust for potential confounders and the impact of other biases. In assessing available epidemiological studies, the findings are usually stratified by type of study design (case report, ecological study, individual-level "case-control" or cohort study) and the ability to adjust for important confounders. At this step of the evaluation, it might be tempting to discount or dismiss findings from case reports as being too subjective. In addition, findings from ecological studies (a type of epidemiological study in which health outcome and exposure information are known only for aggregate populations, not for individuals) might also be dismissed, because these studies are vulnerable to special "ecological biases." Nevertheless, this temptation should be resisted. Case reports and ecological studies have provided important evidence linking chemical exposures to diseases. In addition, a study that appears to use an exposure in an ecologic fashion may avoid the special ecologic biases if exposures are, in effect, imputed for each individual geographical unit (e.g., county, town, or region). This is commonly done in drinking water studies (e.g., Kramer et al., 1992; Bove et al., 1995; Munger et al., 1997).

Although differences in study design and ability to control important confounders may be sources of heterogeneity among studies, the most important and likely sources of heterogeneity in environmental (and occupational) studies are caused by differences in the characterizations of exposures and disease outcomes. Therefore, describing these sources of heterogeneity should be the major focus of the narrative.

On the exposure side, studies may differ in exposure characterization (e.g., yes/no; low, medium, high; exposure based on modeling, sample data, residence, etc.); the level of exposure (a high exposure in one study could be a medium or low exposure in another study, and one study may average the sample data while another uses the maximum value); and the duration, frequency, and timing of exposure (Hertz-Picciotto and Neutra, 1994). Heterogeneity among studies could also be because each study is evaluating an effect at a different point in the exposure-effect curve. In addition, effects (e.g., a particular birth defect) seen at a relatively lower exposure level might differ from effects (e.g., spontaneous abortion) seen at a higher level. While timing of exposure is an issue for adult cancers, it is especially important for outcomes associated with in utero exposures (birth defects, developmental disorders, and childhood cancers).

On the outcome side, studies may differ in the disease grouping evaluated (e.g., cancer grouping by organ system versus subgrouping by histology/grade, all leukemia versus childhood leukemia, etc.). The more etiologically homogeneous the grouping is, the less disease misclassification is introduced and the more likely a true effect will not be underestimated or missed.

To summarize, the narrative of each study/case report should fully discuss potential sources of heterogeneity, with an emphasis on exposure and disease characterization. Although study findings can be summarized in table form by grouping studies according to design, it is often more informative to group studies based on similarity of exposure characterization or exposure level and on similarity of disease characterization.

The narrative and summary tables are key to evaluating the available evidence. However, policymakers usually want some sort of classification framework with criteria in order to judge whether the chemical is likely to be toxic to humans (i.e., to determine whether evidence is sufficient), probably toxic (i.e., human evidence is limited, and in particular, biases cannot be ruled out as explanations for the association), possibly toxic (i.e., human evidence is limited, and there is a lack of supportive evidence from animal studies or case reports), unknown (i.e., no data are available), or possibly or probably not toxic at exposure levels encountered by humans. Candidate criteria that have been included in a proposed framework for the use of epidemiologic studies in risk assessment (Hertz-Picciotto, 1995), a framework used by the Institute of Medicine in its evaluation of herbicide exposure (Mosteller and Colditz, 1996), and a framework instituted by the Nordic Council of Ministers to evaluate the reproductive toxicity of chemicals (Taskinen, 1995) include the following:

- A positive association is present.
- Selection and information (exposure or disease misclassification) biases are reasonably judged as unlikely to account for the positive association (or failure to find a positive association).
- "Chance" is reasonably judged as unlikely to account for the positive association (or failure to find a positive association).
- Confounding bias has been controlled and/or is reasonably judged as unlikely to account for the positive association (or failure to find a positive association).
- Evidence of a (monotonic) dose-response relationship exists.
- The direction of the associations among the studies and with other evidence, including case reports and animal studies, is consistent.

A recent evaluation of studies of alcohol and breast cancer and vasectomy and prostate cancer found that the criteria most often used to assess the evidence for carcinogenesis were (1) strength of the association (as measured, for example, by the risk ratio or the mean difference); (2) consistency across study designs and different populations; (3) existence of a dose-response gradient; (4) biological plausibility; and (5) the impact of biases on the strength of the association (Weed and Gorelic, 1996). While these could be the key criteria for a classification framework, and they correspond with the criteria listed above, the assessment of consistency among studies must take into account the many sources of heterogeneity among studies, especially differences in the characterization of exposures and outcomes.

Judging the impact of chance by referring to an arbitrary standard of statistical significance (e.g., p-value cutoff of 0.05 or the lower limit of the 95 percent confidence interval) is not useful for assessing a study, because it focuses attention on values for the parameter of interest (e.g., the risk ratio) that are not likely (i.e., have very low probability) given the actual results of the study. In addition, whether a study result is statistically significant will depend on the size of the study as well as on the magnitude of the effect. Although a larger study might appear to provide more convincing evidence than a smaller study, it is important to remember that there is often a tradeoff between size and validity. For example, a study may increase its size by diluting both its exposure and disease characterizations and thereby increase the impact of bias.

Predictive Biological Activity or Effect Models

Structure-activity relationship (SAR) and quantitative structure-activity relationship (QSAR) models are used to predict biological activity or effects through the identification of correlations between chemical structure or properties of molecules and biological activities, including those that can be identified through in vitro or in vivo screens and tests. They can be used to predict the biological activity of a number of chemicals, are relatively inexpensive tools, and are most useful when empirical toxicological or epidemiological data are unavailable for specific chemicals within a relatively well-characterized group of related chemicals, such as dioxins.

The SAR approach provides a qualitative means of predicting the hazards of a chemical by developing analogies between chemical substances for which there are few data and chemicals with well-documented health or environmental effects (Lavenhar and Maczka, 1985). The application of QSAR models requires the use of statistical techniques to quantify analogies based on numerical descriptors of physicochemical properties (e.g., lipophilicity, steric parameters, and electronic structure). Describing chemical structures numerically using physicochemical parameters allows the similarity or dissimilarity of a set of compounds to be objectively compared. EPA should systematically review all available SAR/QSAR data and summarize it for use, especially when epidemiological and

---

BOX 5-5   Guiding Principles for Using SAR/QSAR Data in Chemical Priority Setting Efforts (adapted from EDSTAC, 1998)

- The applicable chemical domain of the SAR/QSAR should be as diverse as possible.
- SARs/QSARs should be developed using the most complete and accurate data sets available.
- SARs/QSARs should be validated and used only within the range of conditions for which they are validated.

---

toxicological data are minimal or nonexistent for specific contaminants on the CCL.

The Endocrine Disruptor Screening and Testing Advisory Committee developed guiding principles for evaluating the application of SAR/QSAR models that may be useful to EPA in assessing contaminants on the CCL; Box 5-5 summarizes these principles.

Assessing Treatment Data

Before a final regulatory plan can be established for CCL contaminants, the Safe Drinking Water Act requires that available treatment methods be screened for each contaminant to determine which methods are technologically and economically feasible, which are affordable for small systems, and the degree of risk reduction that can be expected by each of the treatment technologies. The SDWA Amendments of 1986 require EPA to designate a best available technology (BAT) treatment for each contaminant to be regulated. (Designation of a BAT does not require the use of that particular technology to remove the contaminant, but it does require any treatment technique to perform at least as effectively as the BAT.) The SDWA Amendments of 1996 introduced the limited consideration of a cost-benefit analysis in the standard-setting process for certain contaminants. Hence, in establishing a regulation for a contaminant, EPA must determine that a meaningful risk reduction can be achieved by regulating and/or removing that particular contaminant, which means that the performance of treatment technologies must be quantifiable. Another important consideration with respect to regulation and treatment is the size of the public water supply system. The 1996 amendments focus particular attention on this issue. One of the major difficulties in developing and implementing new regulations has been the lack of acceptable and affordable approaches for meeting the needs of small water systems. Larger systems are more likely to have the resources to monitor for specific contaminants on a more frequent basis than do smaller systems, and certain treatment technologies that are feasible for large systems may not be feasible for small systems. Thus, while EPA standards generally are to be set at levels feasible for large systems, the 1996 amendments require EPA to designate acceptable and affordable treatment technologies that can achieve these standards (if any) for small systems, with specific technologies for each of the following service population categories: 25-500; 500-3,300; 3,300-10,000; and greater than 10,000. If no feasible treatment technologies are available for these small systems, variances and exemptions are available.

As for health effects data, EPA should prepare a narrative summary of treatment data for each contaminant on the CCL, but treatment data should not be considered in the preliminary risk assessment recommended in this chapter. The key principle to keep in mind when assessing treatment options for contaminants on the CCL is that the effectiveness of a treatment technology depends on the physical and chemical characteristics of the contaminants in question, and the aquatic matrix in which the contaminants are found. For example, in connection with the existing CCL, contaminants that are only slightly soluble in water (e.g., aldrin, DDE) can be expected to be associated with particles in water and at relatively low dissolved aqueous concentrations. The particulate form of these contaminants should be removable by conventional solid-liquid separation processes, such as coagulation, sedimentation and filtration, and membrane filtration. The dissolved form of these slightly soluble contaminants should be readily removable by such adsorption processes as granular activated carbon adsorption. Contaminants whose solubilities are markedly influenced by pH (e.g., zinc, aluminum) can be removed by first adjusting the pH of the water in which they are found to a level at which they become insoluble, and then removing them by conventional solid-liquid separation processes. If the level of concern is below the solubility limit of the metal, even after pH adjustment, additional treatment processes may be required, such as ion exchange or other chemisorptive processes. For contaminants that are highly polar and have a high aqueous solubility (e.g., perchlorate and some of the substituted phenols), chemical oxidation or reduction, or photolytic or microbial degradation processes may be employed. Membrane processes (e.g., microfiltration, ultrafiltration, and reverse osmosis) if properly staged, are capable of removing both particulate and dissolved contaminants, including such conventional impurities as particulate material and hardness, but these processes can be relatively expensive. In all cases, the technologies that might be implemented to remove these candidate contaminants must do so without interfering with the other objectives of drinking water treatment (e.g., turbidity and color removal, elimination of objectionable tastes and odors) and the removal of other contaminants of health concern. In addition, the recommended processes must not introduce new contaminants to the water that may themselves have an adverse impact on public health.

Assessing Analytical Methods Data

To ensure that data on the occurrence of drinking water contaminants are adequate for exposure assessment, sampling and measurement methods must be reliable and well documented. Analytical methods for currently regulated contaminants in drinking water are well documented and should be adequate for most commonly recognized contaminants that comprise much of the CCL. The greatest analytical challenges lie in the identification of new contaminants and the quantification of emerging contaminants that are intrinsically difficult to measure. Along with written summaries of health effects, exposure, and treatment techniques data, EPA will need to summarize available analytical methods for contaminants on the CCL, focusing especially on newly recognized contaminants.

Chemical Contaminants

From an analytical perspective, it is useful to classify contaminants in drinking water as volatile, semivolatile, and nonvolatile. Volatile organic chemicals (VOCs) have relatively high vapor pressures (0.1 to 380 torr) (Mukund et al., 1995). Therefore, most VOCs are easily purged from the aqueous phase to the gas phase and are separated by gas chromatography. However, if not purged from finished water, remaining VOCs can lead to a large source of exposures by inhalation of indoor air, especially through showering. Variations on this approach have proven to be very robust and are routinely used for the analysis of VOCs in drinking water. In fact, six of the thirteen methods commonly used for determination of organic contaminants in drinking water (see Table 5-2) are for VOCs, and these methods were cited in the *Federal Register* of July 8, 1987, under the National Primary Drinking Water Regulations (EPA, 1987).

In contrast with VOCs, semivolatile organic compounds (SVOCs) have moderate vapor pressures (10^{-7} to 0.1 torr) and are not as amenable to routine analysis. Seven standard methods for non-VOC compounds were cited in proposed drinking water regulations in the *Federal Register* of May 22, 1989, and are also summarized in Table 5-2 (EPA, 1989). However, it will be necessary to develop and standardize new methods for SVOCs and nonvolatile organic compounds in order to obtain the occurrence data necessary to monitor and regulate some of the new and emerging contaminants that may appear on future CCLs. Analytical methods for detecting a wide range of chemical contaminants in drinking water are regularly published in the open research literature; and are not listed in Table 5-2. These methods are not generally validated by EPA, but they represent an important source of information on analytical methods for new and emerging chemical contaminants.

TABLE 5-2   EPA Methods for Determining Organic Compounds in Drinking Water

Number	Method Name
502.1	Volatile Halogenated Organic Compounds in Water by Purge and Trap Gas Chromatography
502.2	Volatile Organic Compounds in Water By Purge and Trap Capillary Column Gas Chromatography with Photoionization and Electrolytic Conductivity Detectors in Series
503.1	Volatile Aromatic and Unsaturated Organic Compounds in Water by Purge and Trap Gas Chromatography
504	1,2-Dibromoethane (EDB) and 1,2-Dibromo-3-Chloropropane (DBCP) in Water by Microextraction and Gas Chromatography
505	Analysis of Organohalide Pesticides and Commercial Polychlorinated Biphenyl Products in Water by Micro-Extraction and Gas Chromatography
507	Determination of Nitrogen- and Phosphorus-Containing Pesticides in Water by Gas Chromatography with a Nitrogen-Phosphorus Detector
508	Determination of Chlorinated Pesticides in Water by Gas Chromatography with an Electron Capture Detector
508A	Screening for Polychlorinated Biphenyls by Perchlorination and Gas Chromatography
515.1	Determination of Chlorinated Acids in Water by Gas Chromatography with an Electron Capture Detector
524.1	Measurement of Purgeable Organic Compounds in Water by Packed Column Gas Chromatography/Mass Spectrometry
524.2	Measurement of Purgeable Organic Compounds in Water by Capillary Column Gas Chromatography/Mass Spectrometry
525.1	Determination of Organic Compounds in Drinking Water by Liquid-Solid Extraction and Capillary Column Gas Chromatography/Mass Spectrometry
531.1	Measurement of N-Methylcarbamoyloximes and N-Methylcarbamates in Water by Direct Aqueous Injection HPLC with Post Column Derivatization

Source: EPA, 1988

Microbiological Contaminants

Methods are available for detecting the presence of almost any microorganism of concern, although difficulties can arise in collecting samples, determining frequency and sample sites, and interpreting the relationship between positive samples and public health (Hurst et al., 1997).

Bacteria

While cultivation techniques are well developed for enteric bacterial indicators, such as coliform and fecal coliform bacteria, little attention has been paid to the development of methods for analyzing enteric bacterial pathogens in water.

This is, in part, because of the historical success of using these indicators in preventing the occurrence of most enteric bacterial waterborne disease outbreaks.

In general, if pathogens are present in great enough concentrations, they can be assayed directly. However, as discussed in Chapter 1 (see Figure 1-1), the causative pathogens of more than half of reported waterborne disease outbreaks are then identified. Three basic methods are used for detection and enumeration of bacteria in environmental samples (Toranzos and McFeters, 1997): (1) most probable number (MPN), (2) membrane filter (MF), and (3) presence-absence (PA).

The MPN method measures the growth of organisms taken from a sample (or a serially diluted sample) on (usually) selective media through production of turbidity, acid, or gas. When the positive tubes have been identified and recorded, it is possible to estimate the number of organisms in the original sample by using an MPN table that gives the number of organisms per certain volume. MPN methods are very labor intensive and require large amounts of media and glassware, and, in the case of pathogens, may require several days to complete. In the MF test, a given volume of liquid is passed through a filter with a pore size less than the diameter of the bacteria, and then the filter is placed on the growth media. The bacteria then grow on the surface of the membrane as individual colonies. This method is more accurate, less time consuming, and more rapid than the MPN method. Lastly, PA tests, while not quantitative per se, can answer the simple question of whether the target organism is present in a sample. Since some standards require the absence of an indicator or pathogen in a certain volume (e.g., 0 coliforms per 100 ml of drinking water), the PA method can be used as a pass/fail screening test.

For the most part, culturable analytical methods have been used for bacteria, however, in some cases, only a small percentage of the total viable organisms present may be detected using these methods of bacterial detection (Colwell et al., 1996). Microscopic techniques, such as the use of antibodies, genetic probes, image analysis, and flow cytometry, have become highly sophisticated, specific, and rapid for the detection of bacteria (Lawrence et al., 1998). Staining with specific genetic probes can address not only total bacterial numbers but the genetic composition and taxonomic status of populations. Thus, the state of the microorganism, as well as its identification, can now be ascertained. Applications for digital microscopy include quantification, viability, metabolic condition, as well as the structure of the microenvironment. However, more emphasis needs to be placed on sample concentration and the use of more specific techniques for bacteria such as *Heliobacter*, which cannot be cultured.

Viruses

Methods for virus detection in water depend on their concentration in volumes ranging from 10 to 2,000 liters. This is accomplished by the adsorption of the viruses to positively charged filters, which adsorb the negatively charged viruses from water (Sobsey and Glass, 1980). Adsorbed viruses are eluted from the filters with a protein solution and further concentrated to a final volume by precipitation of the proteins before assay. These concentrates are assayed in animal cell cultures of human or primate origin. The presence of the virus is indicated by production of cytopathogenic effects (CPE) (the destruction of the individual cells) or formation of plaques or clear zones produced by the destruction of cells under an agar overlay. The isolated viruses are identified by serological neutralization tests.

Currently, viral cultivation methods have largely been optimized for the detection of enteroviruses, and little information is available on other types of viruses that may be present at equal or greater concentrations in drinking water. The filters used to concentrate viruses from water do not concentrate all types of viruses with equal efficiency because of differences in charge on the different types of viruses (Gerba, 1984). Several studies have reported greater concentrations of adenoviruses than enteroviruses (e.g., Grohman et al., 1993) in sewage and sewage-polluted waters. An additional problem is that many viruses (e.g., hepatitis A) may grow in cell culture without the production of CPE. Further, it can take several days to many weeks before the virus produces CPE. A final problem is that sometimes substances are concentrated from the water that are toxic to the cell culture. Additional research is needed to overcome these problems and to develop better techniques for assessing all types of waterborne viruses, not just enteric viruses.

Protozoa

Protozoan parasites are sufficiently large that they can be observed under a normal light microscope, allowing for detection and quantification, and microscopy remains the traditional method for detecting protozoa.

Standard methods have been developed for collection, recovery, and detection of enteric protozoa. Typically, protozoan parasites are collected from large volumes of water by size exclusion through spun filters with a nominal pore size of one micron. These filters also collect suspended matter in the water and this makes visualization of the parasite cysts or oocyst difficult to observe (Rose et al., 1989; LeChevallier and Trok, 1990). To separate parasites from debris, the filters are cut apart and washed with an eluting solution of detergent. The eluate containing the cysts/oocysts and debris is further concentrated by centrifugation, where centrifugation separates the cysts/oocysts from much of the debris. The semi-purified sample is collected from the gradient and labeled with monoclonal antibodies specific to the cyst or oocyst cell wall using a specific immunofluorescent assay (IFA) procedure. The sample can be examined by epifluorescent microscopy for fluorescence, shape, and size, and by phase contrast or Nomarski differential interference contrast microscopy for internal features (LeChevallier et al., 1991a,b).

The efficiency of recovery for cysts/oocysts for this process has been investigated in detail, with overall recovery rates varying from 28 percent to 86 percent for *Giardia* and from 5 percent to 68 percent for *Cryptosporidium* (LeChevallier et al., 1995; Nieminski et al., 1995). However, current methods for recovering and detecting parasites always underestimate the true concentration in environmental samples. While the use of IFA greatly aids the detection of cysts/oocysts, background fluorescence and nonspecific binding of the antibody may decrease their accurate identification. Another limitation is that no single antibody has been found to bind specifically only to species that cause infection in humans; thus, protozoa infecting only lower animals may also be detected. An added problem is that the viability of the cysts/oocysts cannot be assessed by IFA. LeChevallier et al. (1991a) reported that 10 percent to 30 percent of the organisms found in water samples were empty, without internal features, suggesting they were not viable. It is not clear whether this is an artifact of sample processing.

New analytical methods are currently under development for improving both the recovery and detection of protozoa as well as interpretation of the results (Jakubowski et al., 1996). Methods using cell culture infectivity have been successfully applied to address the important question of *Cryptosporidium* viability (Slifko et al., 1997a,b). Immunomagnetic separation (IMS) techniques use antibodies tagged to iron beads and a magnetic system to pull the target oocysts and cysts from the suspension. These techniques have been applied in microscopic detection and polymerase chain reaction approaches (Johnson et al., 1995; Deng et al., 1997). Several IMS kits are now available for *Cryptosporidium* (e.g. Dynal in Lake Success, NY; Crypto-Scan™ in Portland, ME). The use of in-situ hybridization to identify *Cryptosporidium* (Lindquist, 1997) has widespread application for identification and detection efforts, because both microscopy and the specificity of the probe can be used. This also allows for such instrumentation as flow cytometry and digital microscopy to be used, which can greatly reduce the analytical time. To improve understanding of the relationship between potential exposure to waterborne oocysts and cysts and public health outcomes, the greatest research need may be in addressing the viability methods. In addition, when, where, and how often to sample for protozoa should be addressed, with corresponding development of guidance.

Molecular Techniques

Advances in molecular biology have allowed for the development of more rapid, sensitive, and lower-cost approaches to the detection of pathogens in the environment. These methods are designed to detect and analyze the genetic material of the organisms. Since each organism has a unique genetic code, this can be used not only to identify specific species but also to "fingerprint" the strain and clone it. Once a new pathogen has been isolated and its nucleic acid analyzed, methods can be rapidly developed for its detection. These methods also offer the potential to detect microorganisms without the need for cultivation.

The polymerase chain reaction (PCR) has offered the most promise for the rapid detection of pathogens in the environment and has been used for bacteria, protozoa, and viruses (Johnson et al., 1995; Toranzos, 1997). This method involves the specific amplification of the DNA in the genome of the microorganism with the aid of primers. Primers are fragments of DNA that are complementary to the DNA strain to be amplified (sequences specific to the region of the genome to be amplified). Within a few hours, millions of genome copies are produced. The principle of the method involves the repetitive enzymatic synthesis of DNA. Amplification only takes place if the specific nucleic acid of the target organism is present.

PCR has a number of advantages, including (1) specificity of the assay, (2) ability to detect non-cultivable microorganisms, (3) rapidity of the assay (24 hours), (4) ability to conduct multiple assays, and (5) use of automated instrumentation. PCR also has a number of limitations for use directly in environmental samples. First, the maximum volume that currently can be assayed is 0.1 ml. Extracts or concentrates from environmental samples for enteric viruses and protozoa range from 2 ml to 30 ml or more. Thus, further sample concentration is needed (Johnson et al., 1995). Second, environmental samples and concentrates usually contain substances that interfere with detection by masking the target DNA or inhibiting the enzyme reaction. This results in often laborious and time-consuming processing of samples (Abbaszadgan et al., 1993; Schwab et al., 1995; Toranzos, 1997), though it is possible to detect as little as one to two organisms when interfering substances are removed. Lastly, PCR will also detect dead or inactivated microorganisms (Reynolds et al., 1991; Kaucner and Stinear, 1998). Therefore, without cultivation procedures it is not possible to assess viability. While PCR could not be used to assess the performance of disinfection processes, it is still useful for assessing occurrence where viability may not be an immediate need. This is the only method available for the detection of some currently uncultivable waterborne pathogens, such as the Norwalk virus.

SUMMARY: CONCLUSIONS AND RECOMMENDATIONS

In summary, the committee recommends that EPA use a phased process (see Figure 5-1) for determining which contaminants on the CCL are appropriate candidates for regulatory action and which will require research. The recommended process would proceed as follows:

• Within approximately one year of completion of the CCL, EPA should conduct a three-part assessment of each contaminant on the CCL. For each contaminant, the three parts consist of (1) a review of existing health effects data, (2) a review of existing exposure data, and (3) a review of existing data on treatment options and analytical methods. The first part of the assessment should consider data on the contaminant's effects on sensitive populations, such as pregnant women, infants, the elderly, and those with compromised immune systems. While general guidelines for reviewing existing data are possible and are presented in this chapter, an important component of the reviews will be policy judgments by EPA about the significance of the data.

• After completion of the three-part assessment, EPA should conduct a preliminary risk assessment based on available data identified in the three-part assessment. The risk assessment, which integrates hazard and exposure analyses to estimate the public health implications of the contaminant, should be carried out, even if there are data gaps, to provide a basis for an initial decision about the disposition of the contaminant and to guide research efforts, where needed. The preliminary risk assessment, while a critical step in the process, should not be overly detailed or resource intensive.

• After completing the preliminary risk assessment for each contaminant, EPA should prepare a separate decision document, that indicates whether the contaminant will be dropped from the CCL because it does not pose a risk, will be slated for additional research (on health effects, exposure, or risk reduction), or will be considered for regulation. The decision document should explain the reasoning for EPA's determination and should be publicly disseminated for comment. Decision documents for contaminants dropped from the CCL should specify the health and exposure data that EPA used to conclude that the contaminant poses little or no risk.

• When the three-part assessment or preliminary risk assessment identifies important information gaps, EPA should develop a research and monitoring plan to fill such gaps in time to serve as the basis for a revised assessment and decision document before the end of the three-and-a-half-year cycle required by Congress for evaluating contaminants on the CCL. In filling information gaps, EPA should solicit the voluntary participation of industry and others and should use its other authorities (such as those under the Toxic Substances Control Act) to help fill data gaps.

• Health advisories should be issued for all contaminants remaining on the CCL after completion of an initial set of decision documents. A health advisory is an informal technical guidance document that defines a nonregulatory (i.e., nonenforceable) concentration of a drinking water contaminant at which no adverse health effects are anticipated to occur over specific exposure durations. To provide the public with the best available information about the contaminant, EPA should develop a health advisory for any contaminant for which credible evidence of a risk in drinking water exists, even if existing data are insufficient to develop a full regulation. Contaminants subject to a health advisory may need additional research and monitoring even after completion of a revised assessment and decision document.

Decisions to drop a contaminant from the CCL, to issue a health advisory, or to proceed toward regulation should be based on health risk considerations only. However, EPA should fill data gaps in treatment technologies and analytical methods to avoid delaying regulatory action for contaminants for which current information on treatment and detection is inadequate.

In implementing this phased process, EPA should keep in mind that it should act immediately on all contaminants that meet the statutory tests of (1) adversely affecting public health, (2) being known or substantially likely to occur in public water systems with a frequency and at levels that pose a threat to public health, and (3) presenting a meaningful opportunity for health risk reduction. Development of regulations for contaminants that meet these three requirements (which are specified in the SDWA amendments) should not be delayed by implementation of the phased approach. The ability to act quickly and short-circuit the phased evaluation process is especially critical for protecting the public from newly discovered high-risk contaminants. EPA will need to remain flexible in order to be prepared to address such immediate risks.

REFERENCES

Abbaszadegan, M., M. S. Huber, C. P. Gerba, and I. L. Pepper. 1993. Detection of enteroviruses in groundwater with the polymerase chain reaction. Applied Environmental Microbiology 59:1318-1324.

Anderson, B. C. and M. S. Bulgin. 1981. Enteritis caused by cryptosporidiosis in calves. Veterinary Medicine of Small Animal Clinics 76:865-868.

Barker, I. K., and P. L. Carbonell. 1974. *Cryptosporidium agni* sp.n. from lambs and *Cryptosporidium bovis* sp.n. from a calf with observations on the oocyst. Z. Parasitenkd. 44:289.

Bove, F. J., M. C. Fulcomer, J. B. Klotz , et al. 1995. Public drinking water contamination and birth outcomes. American Journal of Epidemiology 141:850-862.

Colwell, R. R., P. Brayton, A. Huq, B. Tall, P. Harrington, and M. Levine. 1996. Viable but nonculturable *vibrio cholera* 1 revert to a cultivable state in the human intestine. World Journal of Microbiology and Biotechnology 12:28-31.

D'Antonio, R. G., R. E. Winn, J. P. Taylor, T. L. Gustafson, W. L. Current, M. M. Rhodes, G. W. Gary, and R. A. Zajac. 1985. A waterborne outbreak of cryptosporidiosis in normal hosts. Annals of Internal Medicine 103:886-888.

Deng, M. Q., D. O. Cliver, and T. W. Mariam. 1997. Immunomagnetic capture PCR to detect viable *Cryptosporidum parvum* oocysts from environmental samples. Applied and Environmental Microbiology 63(8):3134-3138.

EDSTAC (Endocrine Disruptor Screening and Testing Advisory Committee). 1998. Final Report: Volume I. August 1998.

EPA (U.S. Environmental Protection Agency). 1979. National Interim Primary Drinking Water Regulations; Control of Trihalomethanes in Drinking Water; Final Rule. Federal Register 44(231):68624-68707.

EPA. 1984. Aldicarb; Special Review of Pesticide Products Containing Aldicarb. Federal Register 49(134):28320-28323.

EPA. 1987. National Primary and Secondary Drinking Water Regulations; Final Rule. Federal Register 52:25690.

EPA. 1988. Methods for the Determination of Organic Compounds in Drinking Water. EPA-600/4-88/039. Cincinnatti, Ohio: EPA, Environmental Monitoring Systems Laboratory, Office of Research and Development.

EPA. 1989. National Primary and Secondary Drinking Water Regulations; Proposed Rule. Federal Register 54(97):22082.

EPA. 1997. Announcement of the Draft Drinking Water Contaminant Candidate List. Federal Register 62(193):52194-52219.

Federal Focus, Inc. 1996. Principles for Evaluating Epidemiologic Data in Regulatory Risk Assessment. Washington, D.C.

Gerba, C. P. 1984. Recovering viruses from sewage, effluents, and water. In Methods for Recovering Viruses from the Environment, G. Berg, ed. Boca Raton, Fla: CRC Press.

Grohmann, G. S., N. J. Ashbolt, M. S. Genova, G. Logan, P. Cox, and C. S. W. Kueh. 1993. Detection of viruses in coastal and river water systems in Sydney, Australia. Water Science and Technology 27:457-461.

Hayes, E. B., T. D. Matte, T. R. O'Brien, T. W. McKinley, G. S. Logsdon, J. B. Rose, B. P. Ungar, D. M. Word, P. F. Pinsky, M. L. Cummings, M. A. Wilson, E. G. Long, and E. S. Hurwitz. 1989. Large community outbreak of cryptosporidiosis due to contamination of a filtered public water supply. New England Journal of Medicine 320:1372-1376

Hertz-Picciotto, I., and R. R. Neutra. 1994. Resolving discrepancies among studies: The influence of dose on effect size. Epidemiology 5:156-163.

Hertz-Picciotto, I. 1995. Epidemiology and quantitative risk assessment: A bridge from science to policy. American Journal of Public Health 85:484-491.

Hurst, C. J., G. R. Knudsen, M. J. McInerney, L. D. Stetzenbach, and M. V. Walter, eds. 1997. Manual of Environmental Microbiology. Washington, D.C.: ASM Press.

Jakubowski, W., S. Boutros, W. Faber, R. Fayer, W. Ghiorse, M. LeChevallier, J. Rose, S. Schaub, A. Singh, and M. Stewart. 1996. Environmental methods for Cryptosporidium. Journal of the American Water Works Association 88(9):107-121.

Johnson, D. W., N. J. Pieniazek, D. W. Griffin, L. Misener, and J. B. Rose. 1995. Development of a PCR protocol for sensitive detection of Cryptosporidium oocysts in water samples. Applied and Environmental Microbiology 61(11):3849-3855.

Kaucner, C., and T. Stinear. 1998. Sensitive and rapid detection of viable Giardia cysts and Cryptosporidium parvum oocysts in large-volume water samples with wound fiberglass cartridge filters and reverse transcription-PCR. Applied and Environmental Microbiology 64(5): 1743-1749.

Kramer, M. D., C. F. Lynch, P. Isacson, et al. 1992. The association of waterborne chloroform with intrauterine growth retardation. Epidemiology 3:407-413.

Lavenhar, S. R., and C. A. Maczka. 1985. Structure-activity considerations in risk assessment: a simulation study. Journal of Toxicology and Industrial Health 1(4):249-259.

Lawrence, J. R., J. McInerney, and D. A. Stahl. 1998. Analytical imaging and microscopy techniques. Pp. 29-51 (ch. 5) in Manual of Environmental Microbiology, C. J. Hurst, G. R. Knudsen, M. J. McInerney, L. D. Stetzenbach, and M. V. Walter, eds. Washington, D.C.: ASM Press.

LeChevallier M. W., and T. M. Trok. 1990. Comparison of the zinc sulfate and immunofluorescence techniques for detecting Giardia and Cryptosporidium. Journal of the American Water Works Association 82:75-82.

LeChevallier M. W., W. D. Norton, and R. G. Lee. 1991a. Occurrence of Giardia and Cryptosporidium spp. in surface water supplies. Applied and Environmental Microbiology 57(9):2610-2616.

LeChevallier M. W., W. D. Norton, and R. G. Lee. 1991b. Giardia and Cryptosporidium spp. in filtered drinking water supplies. Applied and Environmental Microbiology 57(9): 2617-2621.

LeChevallier M. W., W. D. Norton, J. E. Siegel, and M. Abbaszadegan. 1995. Evaluation of the immunofluorescence procedure for detection of *Giardia* cysts and *Cryptosporidium* oocysts in water. Applied and Environmental Microbiology 61(2):690-697.

Lindquist, J. A. D. 1997. Probes for the specific detection of *Cryptosporidium parvum*. Water Resources 31(10):2668-2671.

MacKenzie, W. R., N. J. Hoxie, M. E. Proctor, S. Gradus, K. A. Blair, D. E. Peterson, J. J. Kazmierczak, K. Fox, D. G. Addiss, J. B. Rose, and J. P. Davis. 1994. Massive waterborne outbreak of *Cryptosporidium* infection associated with a filtered public water supply, Milwaukee, Wisconsin, March and April, 1993. New England Journal of Medicine 331(3):161-167.

Meisel, J. L., et al. 1976. Overwhelming water diarrhea associated with *Cryptosporidium* in an immunosuppressed patient. Gastroenterology 70:1156.

MMWR (Morbidity and Mortality Weekly Report). 1982 Cryptosporidiosis: An assessment of chemotherapy of males with acquired immune deficiency syndrome (AIDS). Morbidity and Mortality Weekly Report 31:589.

Mosteller, F., and G. A. Colditz. 1996. Understanding research synthesis (meta-analysis). Annual Review of Public Health 17:1-23.

Mukund, R., and T. J. Kelly, et al. 1995. Status of ambient measurement methods for hazardous air pollutants. Environmental Science and Technology 29(4):183A-187A.

Munger, R., P. Isacson, S. Hu, et al. 1997. Intrauterine growth retardation in Iowa communities with herbicide-contaminated drinking water supplies. Environmental Health Perspectives 105:308-314.

NRC (National Research Council). 1977. Drinking Water and Health. Washington, D.C.: National Academy Press.

Nieminski, E. C., F. W. Schaefer III, and J. E. Ongerth. 1995. Comparison of two methods for detection of *Giardia* cysts and *Cryptosporidium* oocysts in water. Applied and Environmental Microbiology 61:1714-1719.

Reynolds, K. A., C. P. Gerba, and I. L. Pepper. 1991. Detection of infectious enteroviruses by integrated cell culture-PCR procedure. Applied and Environmental Microbiology 62:1424-1427.

Rose, J. B. 1988. Occurrence and significance of *Cryptosporidium* in water. Journal of the American Water Works Association 80:53-58.

Rose, J. B., L. K. Landeen, R. K. Riley, and C. P. Gerba. 1989. Evaluation of immunofluorescence techniques for detection of *Cryptosporidium* oocysts and *Giardia* cysts from environmental samples. Applied and Environmental Microbiology 55:3189-3195.

Schwab, K. J., R. DeLeon, and M. D. Sobsey. 1995. Concentration and purification of beef extract mock eluates from water samples for the detection of enteroviruses, hepatitis A virus, and Norwalk virus by RT-PCR. Applied and Environmental Microbiology 61:531-537.

Shepard, T. 1994. "Proof" of human teratogenicity. Letter to the editor. Teratology 50:97-98.

Slifko, T. R., D. E. Friedman, J. B. Rose, S. J. Upton, and W. Jakubowski. 1997a. An in-vitro method for detection of infectious *Cryptosporidium* oocysts using cell culture. Applied and Environmental Microbiology 63(9):3669-3675.

Slifko, T. R., D. E. Friedman, J. B. Rose, S. J. Upton, W. Jakubowski. 1997b. Unique cultural methods used to detect viable *Cryptosporidium parvum* oocysts in environmental samples. Water Science and Technology 35(11-12):363-368.

Smith, H. V., and J. B. Rose. 1998. Waterborne cryptosporidiosis: Current status. Parasitology Today 14 (1):14-22.

Sobsey, M. D., and J. S. Glass. 1980. Poliovirus concentration from tap water with electropositive adsorbent filters. Applied and Environmental Microbiology 40:201-210.

Taskinen, H. K. 1995. Nordic criteria for reproductive toxicity. Journal of Occupational and Environmental Medicine 37:970-973.

Toranzos, G. A. 1997. Environmental Applications of Nucleic Acid Amplification Techniques. Lancaster, Pa.: Technomic Publishers.

Toranzos, G. A., and G. A. McFeters. 1997. Detection of indicator microorganisms in environmental freshwaters and drinking waters. Pp. 184-194 in Manual of Environmental Microbiology, C. J. Hurst, G. R. Knudsen, M. J. McInerney, L. D. Stetsenbach, and M. V. Water, eds. Washington, D.C.: ASM Press.

Tyzzer, E. E. 1907. A sporozoan found in the peptic glands of the common mouse. Proceedings of the Society of Experimental Biological Medicine 5:12.

Tzipori, S. 1983. Cryptosporidiosis in animals and humans. Microbiology Reviews 47:84.

Wartenberg, D., and R. Simon. 1995. Comment: Integrating epidemiologic data into risk assessment. American Journal of Public Health 85:491-493.

Weed, D. L., and L. S. Gorelic. 1996. The practice of causal inference in cancer epidemiology. Cancer Epidemiology, Biomarkers, and Prevention 5:303-311.

Westrick, J. J. 1990. National surveys of volatile organic compounds in ground and surface waters. Pp. 103-125 (ch.7) in Significance and Treatment of Volatile Organic Compounds in Water Supplies, N. M. Ram, R. F. Christman, and K. P. Cantor, eds. Chelsea, Mich: Lewis Publishers, Inc.

APPENDIX A

Assessing Uncertainty in Decision Processes

The object of regulating drinking water contaminants is to reduce preventable disease, disability, and death related to drinking water. This assumes that removing, reducing, or preventing contamination of drinking water will result in a mitigation of adverse health effects. Created under the authority of the Safe Drinking Water Act (SDWA) Amendments of 1996, a Drinking Water Contaminant Candidate List (CCL) is a list of contaminants from which decisions to take regulatory action will primarily begin. The framework presented suggests a general scheme EPA might use to develop a decision process to place chemical or microbiological contaminants on a track for regulation, increased research, or removal from a CCL. Realization of this or any other framework requires specific choices that affect how widely or narrowly "the net" will be cast in capturing contaminants that potentially are of public health importance.

As noted in Chapter 5, an ideal decision process is one that exactly selects only those contaminants whose regulation will reduce disease, disability or death, and exactly dismisses those contaminants that play little or no part in affecting human health. Unfortunately, the true state of nature ("the truth") remains either unknown or shrouded in uncertainty for the majority of contaminants on the CCL. It is likely, therefore, that there will be some error in the decision process, allowing some contaminants that should be regulated to pass through, while placing other relatively harmless contaminants on a regulatory track. Thus, there are two kinds of errors that participants in a decision process can make.

Assume there are N contaminants on the list (for the CCL, N = 60, however, it will be demonstrated that the size of N is irrelevant) and EPA uses a decision process that correctly identifies a proportion (s_1) of the contaminants that need regulating (thus leaving $[1 - s_1]$ of these contaminants unidentified) and correctly identifies another proportion (s_2) of contaminants that are harmless. The numbers s_1 and s_2 are features of a particular decision process and will change as the criteria in this process change (e.g., if the agency decides to use a 10^{-6} rather than a 10^{-5} risk level as a trigger for regulation). Assume further that a proportion, p, of contaminants on the list is "truly" in need of regulation.

Given any decision process, it is possible to cross-classify the results in the following 2 × 2 table:

EPA Decision

		Regulate	Do not regulate	
"Truth"	Should regulate	$s_1{*}p{*}N$	$(1 - s_1){*}p{*}N$	$p{*}N$
	Should not regulate	$(1 - s_2)(1 - p){*}N$	$s_2{*}(1 - p){*}N$	$(1 - p){*}N$
		$s_1{*}p{*}N +$ $(1 - s_2)(1 - p){*}N$	$(1 - s_1){*}p{*}N +$ $s_2{*}(1 - p){*}N$	N

It is now possible to answer two important questions: given a decision to regulate, what is the chance that this decision was correct? Given a decision not to regulate, what is the chance that this decision was correct? In screening parlance, the first is commonly called the positive predictive value (PV+), the latter the negative predictive value (PV−), while s_1 is the sensitivity of the decision process, and s_2 is its specificity. The prevalence of the condition that is being screened is p, in this case, the proportion of contaminants on the CCL that should be regulated. Thus, p depends on the criteria used to construct the CCL.

PV+ measures the chance that a contaminant that was selected for regulation was chosen correctly, so $(1 - PV^+)$ represents the fraction of contaminants that were incorrectly selected for regulation. This type of error represents a cost to utilities and the public. PV−, on the other hand, is the fraction of contaminants that are correctly unregulated, so $(1 - PV^-)$ represents the fraction of contaminants that are not regulated but should be. This error represents a public health cost. It can readily be seen that policies affecting PV+ and PV− will have implications for costs and for public health. What determines these two quantities?

To calculate PV+ and PV−:

$$PV^+ = \frac{s_1 pN}{s_1 pN + (1-s_2)(1-p)N} = \frac{s_1}{s_1 + (1-s_2)\frac{(1-p)}{p}}$$

$$PV^- = \frac{s_2 N}{s_2 N + (1-s_1)\frac{p}{(1-p)}N} = \frac{s_2}{s_2 + (1-s_1)\frac{p}{(1-p)}}$$

Note, N is present in every factor in the numerator and denominator and, hence, drops out. Moreover, both PV^+ and PV^- depend on s_1 and s_2 (this is to be expected, as these numbers represent how well a specific decision process performs), but also on p, the proportion of contaminants on the list that need regulating. EPA must contend with a relatively higher p for a list like the CCL, which, by virtue of the way it was developed, is "enriched" with contaminants that likely need regulating relative to a random sample of environmental contaminants.

P is provided by nature, although it can be altered changing the criteria that govern entry onto the candidate list. Given the existing CCL, p is fixed, but s_1 and s_2 can still be adjusted by using different decision processes. Either one of s_1 and s_2 can always be made 100 percent merely by deciding to regulate all contaminants on the list (i.e., s_1 = 100 percent or no contaminants on the list (i.e., s_2 = 100 percent). In the first case, the process correctly identifies all contaminants that need regulating, but will likely drag along many that do not. The reverse is true for s_2 = 100 percent. In some (rare) instances you might have a decision process that has both s_1 and s_2 equaling 100 percent. In that case, one could decide with certainty, in every instance, whether a chemical needs regulating or not. While this is the ideal, it is not the usual circumstance. In most cases, both s_1 and s_2 will fall short (sometimes far short) of the ideal. It is important to examine the consequence of this.

To begin, it is useful to select a few examples.

a) $p = 0.8$ s_1= 90%, s_2 = 90%.

This represents a list with a high percentage of contaminants that truly need regulating, and a decision process that is extremely accurate. It correctly selects 90 percent of the contaminants that need regulating, and ignores 90 percent of those that do not. We calculate PV^+ and PV^-:

PV^+ = 97% PV^- = 69%

Thus, there is little monetary cost to utilities and consumers, but almost one in three (31%) contaminants that should have been regulated were not.

b)   p = 0.5 s_1 = 90%, s_2= 90%.

PV⁺ = 90%   PV⁻ = 90%

In this example, the list is evenly split between contaminants that need regulating and those that do not. The performance of the decision process mirrors s_1 and s_2.

c)   p = 0.1 s_1= 90%, s_2 = 90%.

PV⁺ = 50%   PV⁻ = 99%

While the public health efficiency is low, the utility and consumer efficiency is high, with only 50 percent of the regulated contaminants needing regulation. This represents a list with relatively few contaminants that need regulating, but those that do need regulating are identified.

d)   p = 0.01 s_1= 90%, s_2 = 90%.

PV⁺ = 8%   PV⁻ = 99+%

In the last example, few contaminants on the list actually require regulation, but they are identified. However, the cost to utilities and consumers is high (92 percent of the regulated contaminants do not actually need regulating).

It is important to emphasize that in each example, exactly the same decision process is being used, only the list it operates on is different. It is possible to make some general statements about the effects of different decision processes when p varies. First, both measures of "correctness" depend on s_1, s_2, and p. For high prevalence values, most decision processes (i.e., almost all combinations of s_1 and s_2) lead to low PV⁻ (public health costs in terms of mistaken decisions). On the other hand, for low values of p, most decision processes (combinations of s_1 and s_2) lead to a high proportion of public health decisions being correct. The reverse situation will hold for PV⁺ (a measure of wasted monetary costs).

The criteria used in a given specific decision process determine both s_1 and s_2 and different processes can have different combinations of s_1 and s_2. Generally speaking, a very poor decision process will have low parameters; a good one will have high parameters. But for a given process with criteria that can be adjusted (e.g., how many species of animals are necessary to consider a chemical a carcinogen), changes in the criteria meant to increase either s_1 or s_2 will usually act to decrease the other. The way s_1 and s_2 co-vary can have an important effect on their impact on PV⁺ and PV⁻. To illustrate, consider three generic cases of inverse variation of s_2 with s_1: a linear decrease, a supralinear decrease, and a sublinear decrease (see Figure A-1).

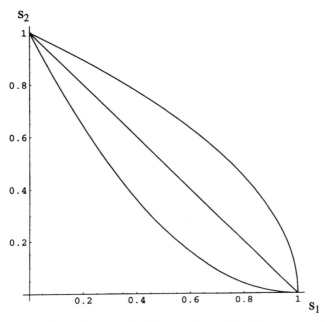

FIGURE A-1   Supralinear, linear, and sublinear relationships between s_1 and s_2.

Linear:	$s_2 = (1 - s_1)$
Supralinear:	$s_2 = (1 - s_1)^{1/2}$
Sublinear:	$s_2 = (1 - s_1)^2$

In all three instances s_2 decreases as s_1 increases (and vice versa), but the effect on PV^- is different in each case (see Figure A-2). Similar effects are produced on PV^+ for different patterns of dependence of s_1 and s_2. What is most surprising is that for some dependencies of s_2 on s_1, both measures can move in the same direction (both up or both down).

Any attempt to increase the proportion of contaminants "captured" by changing decision criteria (e.g., increasing s_1) will have a different effect depending on the specific way that s_2 trades off with s_1. In one case PV^- will increase, in another it will not change, and in a third instance it will decrease. The effect is produced by the speed that s_2 changes with respect to s_1. Thus, the consequences of changing decision criteria can be complex and unpredictable.

Some generalizations are possible. If it is desirable to have both PV^+ and PV^- to be greater than 50 percent (i.e., correct more often than not) then $s_1 + s_2$ ≥ 1.0. However, $s_1 + s_2 \geq 1.0$ only guarantees that PV^+ and $PV^- \geq 1.0$, not that each individually will be ≥ 0.5. When using contaminant lists with a high preva-

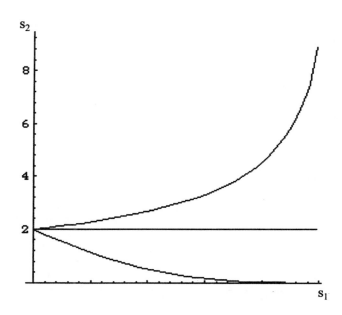

FIGURE A-2   Changes to PV⁻ as decision process altered to increase s_1 as s_2 decreases linearly (flat line), supralinearly (increasing line), or sublinearly (decreasing line).

lence of contaminants that need regulation (like the CCL), a high s_2 is likely to influence PV⁺ (monetary cost efficiency), while high s_1 is more likely to influence PV⁻ favorably (public health efficiency). The reverse is true for lists with an expected low proportion of regulated contaminants (e.g., a list generated with a "wide net" with respect to potentially dangerous contaminants). However, when adjustments are made in the decision criteria used to accommodate desirable performance, it is possible for an unexpected result to appear, depending on the mutual dependence of s_2 and s_1, as shown above.

   In order to predict and evaluate the effects of changing criteria, EPA should consider estimating s_1 and s_2 by applying any contemplated decision process to a group of contaminants that are currently widely accepted as appropriately regulated and a group of contaminants widely accepted as not needing regulation. By changing the criteria to affect s_1 (or s_2) it would also be possible, in principle, to estimate the functional relationship between s_1 and s_2. These general observations are applicable to any decision process, whether applied to elements of the CCL or to general contaminants considered as candidates for the CCL. The analysis demonstrates that the same decision process can produce quite different costs and benefits in these two applications.

APPENDIX B

Biographical Sketches of Committee Members and Staff

Warren R. Muir, Chair, is president of Hampshire Research Institute, Inc., a nonprofit organization in Alexandria, Virginia, and Hampshire Research Associates, Inc., a scientific and engineering consulting firm. Both organizations study issues relating to pollution prevention, risk assessment, and the use of data and information to promote environmental goals. He has held positions as senior staff member for environmental health for the Executive Office of the President, Council on Environmental Quality; deputy assistant administrator for testing and evaluation at the U.S. Environmental Protection Agency; and director of EPA's Office of Toxic Substances. Dr. Muir chaired the NRC Toxicology Information Committee, currently serves as a member of BEST, and has served on several other committees. He received a B.A. in chemistry from Amherst College and M.S. and Ph.D. in chemistry from Northwestern University.

R. Rhodes Trussell, Vice-Chair, is the lead drinking water technologist and director for corporate development at Montgomery Watson, Inc. Dr. Trussell serves on the EPA Science Advisory Board's Committee on Drinking Water. He has served on several NRC committees and is a member of the National Academy of Engineering. Dr. Trussell received his B.S. in civil engineering and his M.S. and Ph.D. in sanitary engineering from the University of California, Berkeley.

Frank J. Bove is a senior epidemiologist for the Epidemiology and Surveillance Branch of the Division of Health Studies, Agency for Toxic Substances and Disease Registry. Dr. Bove has published several papers and reports on the epidemiology of exposure to drinking water contaminants and related adverse

109

health effects. He received a B.A. in political science and philosophy from the University of Pennsylvania, an M.S. in environmental health science, and a Sc.D. in epidemiology from the Harvard School of Public Health.

Lawrence J. Fischer is a professor in the Department of Pharmacology and Toxicology and is the director of the Institute for Environmental Toxicology at Michigan State University. His primary research interest is biochemical toxicology. Specific research includes absorption, distribution, metabolism, and excretion of drugs and chemicals and toxicity of chemicals to the endocrine pancreas (gland). Dr. Fischer received his B.S. and M.S. in pharmacology from the University of Illinois and his Ph.D. in pharmaceutical chemistry from the University of California, San Francisco.

Walter Giger is a professor and senior scientist in the Chemistry Department at the Swiss Federal Institute of Environmental Science and Technology. His research, teaching, and consulting activities focus on organic compounds in the environment and in the geosphere. Research topics include development of analytical techniques for identification of organic pollutants in drinking water, wastewater, and natural waters; investigation of sources, occurrences, and fate of organic pollutants in wastewater and drinking water; and evaluation of chemical, physical, and biological processes that determine the environmental fate of chemicals. Dr. Giger received his B.S. and Ph.D. in chemistry from ETH Zurich.

Branden B. Johnson is a research scientist in the Division of Science and Research at the New Jersey Department of Environmental Protection. His research interests and work activities include broad areas of risk communication, risk perception, natural and technological hazard management, and environmental policy. Dr. Johnson is currently involved in research related to the Consumer Confidence Report requirements of the Safe Drinking Water Act Amendments of 1996 and public reaction to information on *Cryptosporidium* in drinking water. He received a B.A. in environmental values and behavior from the University of Hawaii, Manoa, an M.A. in environmental affairs (water resources), and a Ph.D. in geography from Clark University.

Nancy K. Kim is director of the Division of Environmental Health Assessment of the New York State Department of Health and is an associate professor in the School of Public Health at the State University of New York, Albany. Her research interests include chemical risk assessment, exposure assessment, toxicological evaluations, structural activity relationships, and quantitative relationships among toxicological parameters. She received her B.A. in chemistry from the University of Delaware and her M.S. and Ph.D. in chemistry from Northwestern University.

Michael J. McGuire is president and founder of McGuire Environmental Consultants, Inc., in Santa Monica, California. The firm provides consulting services to public water utilities and industries in the areas of Safe Drinking Water Act compliance and water treatment optimization. Prior to forming his own corporation, he was assistant general manager of the Metropolitan Water District of Southern California. Dr. McGuire received his B.S. in civil engineering from the University of Pennsylvania and his M.S. and Ph.D. in environmental engineering from Drexel University.

David M. Ozonoff is a professor in and chair of the Department of Environmental Health in Boston University's School of Public Health. His research work centers on health effects to communities resulting from various kinds of exposures to toxic chemicals; new approaches to understanding the results of small case-control studies; and the effects of exposure misclassification in environmental epidemiology. He has studied public health effects resulting from exposure to a number of contaminated sites. Dr. Ozonoff received his M.D. from Cornell University in 1967 and his M.P.H. from The Johns Hopkins School of Hygiene and Public Health in 1968.

Catherine A. Peters is an assistant professor in the Program of Environmental Engineering and Water Resources in the Department of Civil Engineering and Operations Research at Princeton University. Her research interests include the behavior of multicomponent organic contaminants in the environment, with particular emphasis on non-aqueous phase liquids (NAPLs); innovative mathematical modeling approaches for characterization of chemical heterogeneity of pollutants; and risk-based decision making for complex multicomponent contaminants. She received her B.S.E. in chemical engineering from the University of Michigan and her M.S. in civil engineering and Ph.D. in civil engineering/engineering and public policy from Carnegie Mellon University.

Joan B. Rose is a professor in the Marine Science Department at the University of South Florida. Her research interests include methods for detection of pathogens in wastewater and the environment; water treatment for removal of pathogens; wastewater reuse; and occurrence of viruses and parasites in wastewater sludge. Dr. Rose served on NRC's Committee on Wastewater Management for Coastal Urban Areas and the Committee on Potable Water Reuse. She received a B.S. in microbiology from the University of Arizona, an M.S. in microbiology from the University of Wyoming, and a Ph.D. in microbiology from the University of Arizona.

Philip C. Singer is a professor in and director of the Water Resources Engineering Program at the University of North Carolina, Chapel Hill. Dr. Singer was formerly a member of NRC's Water Science and Technology Board and served on the Committee on U.S. Geological Survey Water Resources Research. A member of the National Academy of Engineering, he has published dozens of papers and reports principally concerned with aspects of water chemistry and drinking water quality. He received his M.S. and Ph.D. in environmental sciences and engineering from Harvard University.

Deborah L. Swackhamer is an associate professor in the Division of Environmental and Occupational Health in the School of Public Health at the University of Minnesota. Her research involves assessment of contaminants in the environment and associated risks to public health and the environment. She has published dozens of papers on topics ranging from inventories of xenobiotic organic compounds in the Great Lakes, to analytical methods for contaminant detection, to bioaccumulation of organochlorine compounds in fish and multimedia approaches for modeling human exposure. She has served on the executive committee of the Division of Environmental Chemistry of the American Chemical Society, the Board of Directors of the International Association for Great Lakes Research, and the Science Advisory Committee of EPA's Great Waters program. She was a member of the National Research Council's Committee on Coastal Oceans. Dr. Swackhamer received her M.S. in water chemistry and her Ph.D. in oceanography and limnology from the University of Wisconsin, Madison.

Paul G. Tratnyek is an associate professor in the Department of Environmental Science and Engineering and the Department of Biochemistry and Molecular Biology at the Oregon Graduate Institute of Science and Technology. He is also an affiliated scientist with the Center for Coastal and Land-Margin Research and the Center for Groundwater Research. His research primarily involves a wide range of oxidation-reduction reactions that can occur in the environment and the contribution of these reactions to the fate of organic pollutants. Examples include oxidations by chlorine dioxide and oxidations of gasoline oxygenates, such as MTBE. Dr. Tratnyek received his B.A. in chemistry from Williams College and his Ph.D. in applied chemistry from the Colorado School of Mines.

STAFF

Jacqueline A. MacDonald is associate director of the NRC Water Science and Technology Board. She directed the studies that led to the reports *Innovations in Ground Water and Soil Cleanup; Alternatives for Ground Water Cleanup; In Situ Bioremediation: When Does It Work?*; *Safe Water From Every Tap: Improving Water Service to Small Communities;* and *Freshwater Ecosystems: Revitalizing Educational Programs in Limnology.* She received the 1996 National Research Council Award for Distinguished Service. Ms. MacDonald earned an M.S. degree in environmental science in civil engineering from the University of Illinois, where she received a university graduate fellowship and an Avery Brundage scholarship, and a B.S. degree magna cum laude in mathematics from Bryn Mawr College.

Carol A. Maczka, is the director of toxicology and risk assessment at the NRC Board on Environmental Studies and Toxicology. She received her Ph.D. in pharmacology from the George Washington University, with a minor in the metabolism of xenobiotics. She received a B.A. from the State University of New York at Stony Brook in 1973. Dr. Maczka directed the studies that led to the following reports: *Assessment of Exposure-Response Functions for Rocket-Emission Toxicants; Toxicological and Performance Aspects of Oxygenated Motor Vehicle Fuels; and Nitrates and Nitrites in Drinking Water.* Other current projects include: Arsenic in Drinking Water; Hormonally Active Agents in the Environment; Developmental Toxicology; and Strategies To Protect the Health of Deployed U.S. Forces.

Mark C. Gibson is a research associate at the NRC Water Science and Technology Board. He received his B.S. in biology from Virginia Polytechnic Institute and State University and M.S. in environmental science and policy in biology from George Mason University. Mr. Gibson helped form the committee, schedule meetings, and prepare and edit this report.

Kimberly A. Swartz is a project assistant with the NRC Water Science and Technology Board. She assisted the staff and committee in producing the final draft of this report. She has a B.S. in sociology from Virginia Polytechnic Institute and State University.